PERSON-CENTERED STUDIES IN PSYCHOLOGY OF SCIENCE

This unique collection examines "the acting person" as an important unit of analysis for science studies, using an integrative approach of in-depth case studies to explore the cognitive, social, cultural, and personal dimensions of a series of key figures in the sciences, from Goethe to Kepler to Rachel Carson.

Opening up key questions about what science is, and what comprises a scientist, the volume offers an accessible introductory approach to psychology of science, a growing area in Science and Technology Studies (STS). Case studies focus on the psychological contexts of the contributions for which the scientist is known. Without diminishing its epistemic authority, science is presented as a psychologically saturated human activity, one that is especially illustrative of the way social, cognitive, and personal processes intermingle to both facilitate and impede scientific accomplishment. Each case study ends with a set of discussion questions, providing a valuable resource for student reflection and discussion, inviting analysis of similarities and differences in science in the context of very different lives and different projects.

Person-Centered Studies in Psychology of Science is essential reading for scholars and graduates interested in the psychology of science, personality theory, social, or cognitive psychology, general psychologists, and theoretical psychologists.

Lisa M. Osbeck is Professor of Psychology at the University of West Georgia, Fellow of the American Psychological Association, and a past-president of the Society for General Psychology (APA Division 1).

Stephen L. Antczak's academic publishing credits feature in the publications *Qualitative Psychology*, the *Journal of Constructivist Psychology*, and the *Journal of Theoretical and Philosophical Psychology*. Areas of interest include the communication of knowledge in science teams and the subjective experiences of working scientists.

"*Person-Centered Studies in Psychology of Science* maintains that various psychological features of individual scientists are indispensable for a thoroughgoing understanding of science itself. 'Case studies' of prominent historically-and-disciplinarily diverse scientists vividly illustrate this indisputable-yet-overlooked fact. Taken as a whole, the book provides a missing link in contemporary science studies: How appreciating the personal, cognitive, and social dimensions of scientists themselves sheds new light on their well-known contributions to knowledge about the world and ourselves. Students and educators will find much to discuss in these pages, written by authors who focus their psychological lenses on the workings of science at the level of scientists themselves."

Barbara S. Held, *Barry Wish Professor of Psychology and Social Studies Emerita, Bowdoin College, USA*

"In this fascinating, provocative set of explorations of the lives of particular scientists, the authors shed new light on the psychology of science and on the nature of science itself."

Alan Tjeltveit, *Professor of Psychology Emeritus, Muhlenberg College, USA*

"I recommend this gem of a book to a very broad audience. More than pedagogical, it is a ground-breaking contribution to the psychology of science and to qualitative, person-centered psychology. A welcome addition to undergraduate and graduate curricula that will attract and edify students at all levels, this fascinating collection will also interest the lay public. Its engaging style and substance will be enjoyed by readers as their understanding of both science and psychology are challenged and enhanced."

Frederick J. Wertz, *Professor Emeritus, Fordham University, USA*

PERSON-CENTERED STUDIES IN PSYCHOLOGY OF SCIENCE

Examining the Active Person

Edited by Lisa M. Osbeck and Stephen L. Antczak

Routledge
Taylor & Francis Group

NEW YORK AND LONDON

Designed cover image: Getty

First published 2023
by Routledge
605 Third Avenue, New York, NY 10158

and by Routledge
4 Park Square, Milton Park, Abingdon, Oxon, OX14 4RN

Routledge is an imprint of the Taylor & Francis Group, an informa business

Library of Congress Cataloging-in-Publication Data
Names: Osbeck, Lisa M., 1962– editor. | Antczak, Stephen L., 1966– editor.
Title: Person-centered studies in psychology of science : examining the active person / edited by Lisa M. Osbeck and Stephen L. Antczak.
Description: New York, NY : Routledge, 2023. |
Includes bibliographical references and index.
Identifiers: LCCN 2022030809 (print) | LCCN 2022030810 (ebook) |
ISBN 9781032233086 (paperback) | ISBN 9781032233093 (hardback) |
ISBN 9781003276692 (ebook)
Subjects: LCSH: Scientists–Psychology–Case studies. | Research–Psychological aspects–Case studies. | Science–Psychological aspects–Case studies. |
Science and psychology–Case studies.
Classification: LCC Q180.55.P75 P47 2023 (print) |
LCC Q180.55.P75 (ebook) | DDC 509.2/2–dc23/eng20221013
LC record available at https://lccn.loc.gov/2022030809
LC ebook record available at https://lccn.loc.gov/2022030810

ISBN: 978-1-032-23309-3 (hbk)
ISBN: 978-1-032-23308-6 (pbk)
ISBN: 978-1-003-27669-2 (ebk)

DOI: 10.4324/9781003276692

Typeset in Bembo
by Newgen Publishing UK

CONTENTS

CONTRIBUTORS

Stephen L. Antczak's (MLS, ABD) academic publishing credits feature in the publications *Qualitative Psychology*, the *Journal of Constructivist Psychology*, and the *Journal of Theoretical and Philosophical Psychology*. Areas of interest include the communication of knowledge in science teams and the subjective experiences of working scientists.

Ahmed Asad holds a bachelor's degree in Psychology from Foundation University, Pakistan. He is currently pursuing a Ph.D. in *Consciousness and Society* from the University of West Georgia. His research interests revolve around the dialogue between the philosophy of Charles Peirce and theoretical issues in psychology.

Georgia F. Crowe is currently entering her first year pursuing a doctorate in *Consciousness and Society* at the University of West Georgia. Her primary research interests are in exploring narrative as a meaning-making process. This chapter is her first publication.

Ron C. Hopkins has an M.A. in Psychology from the University of West Georgia and an M.S. in Educational Psychology from the University of Wisconsin-Madison. He is currently working on his doctorate in *Consciousness and Society* from the University of West Georgia.

Muhammad Azam Khalid has an M.D. from Pakistan and is a psychiatrist by training. Currently, he is pursuing a Ph.D. in *Consciousness and Society* from the University of West Georgia. His research interests lie in the intersection of Psychology and the works of Charles Peirce and Josiah Royce.

Ronald B. Miller, Ph.D., is Professor Emeritus of Psychology at Saint Michaels College in Colchester, VT. He is the author of the books *Not So Abnormal Psychology: A Pragmatic View of Mental Illness* (2015), *Facing Human Suffering: Psychology and Psychotherapy as Moral Engagement* (2004), and the editor of *The Restoration of Dialogue: Readings in the Philosophy of Clinical Psychology* (1992) all published by the American Psychological Association Press. Dr. Miller has also been a practicing clinical psychologist for 45 years.

Lisa M. Osbeck is Professor of Psychology at the University of West Georgia. Previous works include *Science as Psychology: Sense-Making and Identity in Science Practice* with N. Nersessian, K. Malone and W. Newstetter (2010, Cambridge,), *Rational Intuition*, with B. Held (2014, Cambridge), *Values in Psychological Science* (2019, Cambridge), and *Psychological Studies of Science and Technology*, with K. O'Doherty, E. Schraube, and J. Yen (2019, Palgrave). She is a fellow of the American Psychological Association and a past-president of the Society for General Psychology (APA Division 1).

Yousaf Raza is a consultant psychiatrist from Pakistan and is currently pursuing his Ph.D. in Psychology from the University of West Georgia, USA. He holds a Diplomate in Logotherapy where he focused on the postmodern relevance of Viktor Frankl. He is interested in developing a dialogue between Pragmatism and Psychoanalysis and in doing so facilitate the reflexive exchange between theory and practice.

Peder Schillemat has an M.A. in Psychology from the University of West Georgia and a B.S. in Psychology from Brigham Young University. He is currently pursuing his doctorate in *Consciousness and Society* from the University of West Georgia.

Michael V. Steder has an M.A. in Psychology from the University of West Georgia. He is currently working on a dissertation for his doctorate in *Consciousness and Society* from the University of West Georgia. Research interests include Jungian psychoanalysis and dissociative disorders.

ACKNOWLEDGMENTS

We gratefully acknowledge our colleagues in the Department of Anthropology, Psychology, and Sociology (DAPS) and the College of Arts, Culture, and Scientific Inquiry at the University of West Georgia. Thanks to three anonymous reviewers for their helpful suggestions on the proposed manuscript. Sincere thanks to the editorial staff in the Social Science Division at Routledge, especially Eleanor Taylor, Tori Sharpe, and Alex Howard, for their kind and supportive guidance through all stages of manuscript preparation and production. We also greatly appreciate Dr. Ronald B. Miller's participation in the project of this book, for his insightful commentary on the chapters and helpful suggestions on organization and content.

INTRODUCTION

Person-Centered Studies in Psychology of Science

Lisa M. Osbeck and Stephen L. Antczak

Overview

This book proceeds from an acknowledgment that science is an important form of human activity that deserves more attention than psychology has given it. What makes science different from everyday sense-making? What is the exact nature of science? These questions remain topics of debate and complicate the matter of how we should regard the products of scientific activity (scientific theories, models, and facts). Clearly, however, whatever answers we might give, we cannot deny that science is made possible in part through the goal-directed activity of persons. As expressed by William James, "[h]uman motives sharpen all our questions, human satisfactions lurk in all our answers, all our formulas have a human twist" (James, 1907/1975, p. 117).

As persons, scientists have meaningful relationships, hopes, fears, insecurities, desires, values, and commitments, none of which they leave behind – though they may try – when they engage in science. Their reasoning involves various forms of logic but also inspiration and imagination, and inevitably occurs through interaction with others, living or dead. Discoveries may have enduring significance, but they are always made in a distinct cultural and historical context that empowers some persons and stifles others. Innovative ideas are positioned in relation to prevailing beliefs and a body of existing knowledge, and at least initially with the instruments and technologies available at that time. The world does not always behave as scientists expect it will, so the best of their ideas may need revision. Yet somehow, under some conditions, scientists are capable of "establishing contact with a hidden reality; a contact that is defined as the condition for anticipating an indeterminate range of yet unknown (and perhaps yet inconceivable) true implications" (Polanyi, 1974/1958, p. viii). The exact workings of science are

DOI: 10.4324/9781003276692-1

thus exceedingly intricate and still somewhat elusive, requiring a multitiered and interdisciplinary approach to understand all that is involved.

Reflecting its complexity, science is studied through different disciplinary lenses. Philosophy, sociology, and history all offer insights, with corresponding subdisciplines (e.g., philosophy of science, Science and Technology Studies – STS) now well established in the academy. Psychology's contribution to the interdisciplinary study of science has been less prominent, in part, because psychology itself is a variegated discipline. Cognitive psychology, personality psychology, and social psychology advance understanding with their own concepts applied to the study of science, but there is very little metatheory that brings these together in a cohesive way (see Feist, 2008; Feist & Gorman, 2013). Diverse methods are also used. Experimental analyses of scientific reasoning (e.g., Schunn & Anderson, 1999; Klahr & Dunbar, 1988), cognitive historical (e.g., Nersessian, 1984; Tweney, 1992) and *in vivo* investigations of scientists (e.g., Dunbar, 2000; Nersessian, 2022), as well as correlational studies of personality traits of scientists (e.g., Eiduson, 1962; Feist, 1998; Simonton, 2004) traditionally represent psychology of science (see also Klahr & Simon, 1999).

More recently, psychologists using qualitative studies of science, feminist theory, and critical theory perspectives have broadened the methodological and conceptual domain of psychology of science into "psychological studies of science and technology" (see O'Doherty, Osbeck, Schraube, & Yen, 2019). In short, psychology of science currently takes many forms, and aims at different levels of generality. Studies may be taken as representative of all people, as is indicated by focus on the developmental progression of scientific thinking in children (e.g., Gopnik, 2012); of scientists as a subset or type of person(s) with discernible characteristics (e.g., Eiduson, 1962); of some group of scientists working together (a working lab, or scientists who identify as "bioengineers," Nersessian, 2022; Osbeck & Nersessian, 2017); or of individual scientists and their lives (see Runyan, 2013).

The particular focus of this book is the study of historical scientists as unique persons and the various psychological contexts of some of the scientific contributions for which they are known. The chapters offer case studies of a wide range of scientists from different sciences and time periods, selected for reasons personal to the chapter authors. The book is intended especially for psychology students, for two principal reasons. The first is that psychology students often are taught that psychology is a science and that they are to be trained as scientists. However, they may receive little opportunity to reflect on what that means, beyond the application of what is promoted as "the scientific method" to psychological phenomena. For these students, we hope the exploration of the psychological side of science will broaden their appreciation of the ways science is conducted, and the many kinds of life events that can contribute to discovery and theory construction. On the other hand, some psychology students retreat from science, finding it bloodless, sterile, or fearing it to be mechanistic or reductionistic. For these students, we hope the book will help them to recognize science as a fascinating, psychologically saturated human activity, one that is especially illustrative of the way social, cognitive, and personal processes intermingle to both facilitate and impede empirical knowledge.

The chapters as a set invite discussion of questions such as *"What is science? What does it mean to do science, and when is one doing it? What is a scientist?"* There are also more specific questions relating to the particular focus and scientist covered in each chapter. In addition to encouraging dialogue on science in general, we hope that these case studies also stimulate reflection on how psychological science may be like and different from other forms of science.

Background and Context

Apart from the introduction and conclusion, the chapters in this book were written by graduate students enrolled in a seminar called "Psychology of Science" at the University of West Georgia, in a psychology program with a long history of commitment to humanistic and, more recently, critical psychology perspectives. The course examined cognitive, social, and personal dimensions of science as a human activity, framed as different but overlapping aspects of the psychological dimension of science. To bring the concepts to life, students were assigned to conduct a case study of a scientist of their choosing, someone whose life or work was personally meaningful and interesting to the student. Their task was to relate course content to the scientist by describing or analyzing the "full psychological context" of the scientist's primary achievements, whether in discovery or application. By full psychological context, the expectation was that social (both broad scale, such as organizational, institutional, or cultural level; and local, or family and collegial relationships), cognitive (forms of reasoning employed), and personal (disposition, skills, habits, commitments) factors would be brought out and discussed in relation to specific scientific contributions. This is what students have done. Their storytelling and analysis were so varied and insightful that we began to think that their studies could be of value to others as pivots for discussion of the human side of science. Each chapter is quite different, bearing the peculiarity of the chapter author as much as the scientist discussed, and focusing on a theme of interest to the author within the broad parameters of the assignment.

As a collection of case studies, the chapters may be a resource for student reflection and discussion, studying science in the context of a set of very different projects and very different lives, and offering opportunity for analysis of similarities and differences across scientists. Individual chapters prompt discussion on topics that include the role of values in science, the impact of a scientist's temperament, the meaning of and lines of division between objectivity and subjectivity, the role of intuition in reasoning, spirituality and science, culture and science, science and activism, feminist critiques of science, and perspectivity. Discussion might be focused on one case study at a time, on common themes and differences across case studies, or by comparison of different pairs of scientists. The book ends with an insightful chapter of commentary by a philosophically astute psychologist well known for theoretical and empirical writings on the use of case studies in psychology (e.g., Miller, 2004, 2011).

Conceptual Framework

Science as Personal Engagement

Carl Rogers, in what he called "a highly personal document," describes his struggle to come to terms with the distance between his roles as scientist and therapist, the one demanding "rigorous objectivity" and the other an "almost mystical subjectivity" (Rogers, 1955, p. 267). Eventually, he reconciles the conflict for himself by offering a "change of emphasis" in his description of science, one that foregrounds subjective processes in every step, from the creative inception leading to discovery to the broad social use of science:

> Science exists only in people. Each scientific project has its creative inception, its process, and its tentative conclusion, in a person or persons. Knowledge – even scientific knowledge – is that which is subjectively acceptable. Scientific knowledge can be communicated only to those who are subjectively ready to receive its communication. The utilization of science also occurs only through people who are in pursuit of values which have meaning for them.
>
> *Rogers, 1955, p. 274*

In the following decade, Abraham Maslow offered a longer, more detailed discussion which he framed as a psychology of science, with the goal of attempting to "enlarge" and "rehumanize" science (Maslow, 1966, p. xvi). The focus of his critique of the received view of science is its supposition of value neutrality, as well as the detached stance he viewed as contrary to his own experience of practicing science as a form of engaged encounter. Maslow also depicts science as essentially a matter of "keeping in touch with reality, keeping your eyes open," which he calls "almost a defining characteristic of humanness" (Maslow, 1966, p. 135), With different methods and frameworks, similar points are raised by psychologist Ian Mitroff, in a superb analysis of interviews with NASA scientists (Mitroff, 1974), and Michael Mahoney, in strongly countering a view of science as disinterested neutrality and argued instead for the infusion of emotion and ambition in all phases and levels of scientific practice (Mahoney, 1976, 1979).

The rich psychological analyses of these authors overlap with and implicate philosophical assumptions concerning the nature of knowledge, especially the relation of the knower to the known. An earlier source explicitly acknowledged as an important influence by Maslow, at least, is chemist and philosopher Michael Polanyi's *Personal Knowledge* (1974/1958). Polanyi calls scientific detachment a "false ideal" and offers "conceptual reform" (p. vii) rooted in analysis of the central role of values, passions, aesthetic sensibilities, commitment, skill, and dispositional qualities suitable for the forms of collaboration required of scientists. At the same time, with finely analyzed examples from physical science, Polanyi argues that "the act of knowing includes an appraisal," calling this a "personal coefficient, which shapes all factual knowledge" (1974/1958, p. 17). The passionate striving entails a commitment, and intellectual commitment is infused with passion. On Polanyi's view, the active nature of commitment distinguishes it from other emotions that are more passive, such as purely sensual pleasure. For this reason, he proposes a distinction between "the personal," which he describes as something that "actively enters into our commitments," and subjective states, "in which we merely endure our feelings" (Polanyi, 1974/1958, p. 300). Moreover, commitment "bridges in doing so the disjunction between subjectivity and objectivity," for the only means of transcending one's subjectivity is by "striving passionately" to meet one's "personal obligations to universal standards" (Polanyi, 1974/1958, p. 17).

> In so far as the personal submits to requirements acknowledged by itself as independent of itself, it is not subjective; but in so far as it is an action guided by individual passions, it is not objective either. It transcends the distinction between subjective and objective.
>
> *Polanyi, 1974/1958, p. 300*

Persons

We identify the chapters in this book as "person-centered" studies of science, meaning that their focus is on the active (acting) person of science, whose passions and commitments, skills and aesthetic responding are inevitably, irreducibly involved in the production of scientific knowledge. This does not mean that "the personal" is all that is involved, that science is reducible to it. There are logical requirements, cognitive operations, social conventions and interactions, material constraints, economic and other opportunities, all of which contribute to the intricate mix. Our conviction, however, is that a psychology of science that does not include "the personal" is not complete.

"The personal," however, should be understood against an understanding of "the person," which we have not yet provided. The nature of "person" as an ontological kind is a longstanding philosophical problem, as Rogers acknowledges in his attempt to reconcile what he initially saw as competing demands of personhood and science (Rogers, 1955). A subset of psychologists, Rogers among them, have wrestled with the implications of the philosophical person for a discipline supposedly concerned with them. We can best appreciate the meanings associated with "person" by considering the contrasts to a psychology centered on the person. For early psychologist William Stern, whose careful thought is illuminated by contemporary theorist James Lamiell, a person as a "unitary, self-activated, goal-directed being," a *unitas multiplex*, contrasts with a *thing* (Stern,

1906, p. 16, quoted from Lamiell, 2012, p. 379; See also Lamiell, 2021). Rogers's "person-centered" approach to therapy, providing an environment of freedom, empathy, and acceptance within which the client may explore and grow, contrasts with a problem-centered approach more typical in its time (Rogers, 1961; see also Schmid, 1998). Contemporary theoretical psychologists who advocate greater attention to "the person" contrast a person-centered approach with a psychology focused on either biological or cognitive mechanism or, equally problematic, one focusing exclusively on sociocultural relations (see Smythe, 1998). For Bergman and Andersson (2010) the contrast is between persons and variables. Lundh provides an excellent review of several meanings of person-focused research in psychology (Lundh, 2015), and positions emphasis on the person as a corrective to the various crises (e.g., replication) currently plaguing psychology.

For Svend Brinkmann, "personism" contrasts with "mindism" and "brainism," on the basis that

> [t]o be a person is to find oneself situated in a normative framework, confronted with various rights, duties, and reasons for action … This means that we do not ascribe personhood to creatures on the basis of inner minds or brains, but rather on the basis of their abilities to act, engage in self-reflection, and live up to obligations.
>
> *Brinkmann, 2021, p. 23*

Mark Bickhard rejects the idea of person as a metaphysical kind but acknowledges that persons defy operationalization. He forwards a model of persons as a form of process, complex agents fully participating in a social world. His conclusion positions persons as the "loci" of the psychological and conceptually central to the discipline as we *should* conceive it: "Persons are at the center of what psychology studies; persons are the loci of psychological phenomena. Psychology needs to change its conceptions of science and the philosophy of science so that it can take that center into scientific account?" (Bickhard, 2017, p. 5).

Jack Martin writes extensively on the nature of person and regards focus on the person as key to a psychology that avoids reductive blunders. For example, in what he recently calls "a person-based ontology" requires seven components for a minimally adequate definition of "person," for each of which he provides discussion and analysis: "Persons are (1) embodied, (2) self-interpreting, (3) human, (4) agents, (5) with a distinctive ontology, (6) unique capabilities and dispositions, and (7) moral, existential concerns" (Martin, 2021, p. 392). These few examples should suffice to demonstrate the recent attention to the importance of "person" as central to psychological inquiry, description, and theory, and the idea that such a focus will overcome the shortcomings of more reductive approaches. They also illustrate the variety, complexity, and philosophical overlay of definitions and discussions of persons. Extracting across these, we can identify uniqueness, agency, meaning-making, and participation in social/linguistic communities as common to conceptions of person, and integrative, holistic accounts as essential to person-centered studies.

The emphasis on participation, relatedness, and responsibility helps in clarifying why a "person-centered" psychology is not equivalent to an atomistic individualism that has been the well-deserved target of disciplinary critique (e.g., Reber & Slife, 2021). Brinkmann, for example, following Harré (1984) and Sprague (1999), posits *inter*action as key to understanding persons. Persons not only act but react to others, to whom they relate as "irreducible sources of action" (Brinkmann, 2021, p. 24). In addition to power to act they also have responsibilities. In relating persons to science, the emphasis on responsibility and interaction underscores the moral dimension of personhood that ultimately inflects science in myriad ways (see Allport, 1947; Clegg, 2022).

In considering the moral dimension of science, a return to Polanyi's concept of "commitment" is also instructive. As noted, Polanyi differentiates the active, goal-directed nature of commitment from the more passive experience of emotion, on which basis he distinguishes between "personal" and "subjective" aspects of human experience and action. Commitment narrows ones focus and set reasoned, if not inviolable, limits on actions. For example, commitment to one's family is felt, experienced, but carries requirements and impacts others; it transcends individual and social categories in this sense. Commitment to a religious tradition, an occupation, or a political cause also focuses, directs, and limits in similar ways. Commitment to a scientific endeavor not only requires commitment to a goal but also relations to other persons and other things (the objects of study) that recognize their action potential and forms of agency (even if not human). Science, that is, entails cooperative participation with other agents; moreover, its products (theories, applications) have consequences for others. We can see, therefore, that the line between commitment and obligation is a fine one: "Like love, to which it is akin, this commitment is a 'shirt of flame,' blazing with passion and also like love consumed by devotion to a universal demand" (Polanyi, 1974/1958, p. 64).

The emphasis on interaction and commitment helps to underscore that agents are not unlimited in their agency; they face restrictions of various kinds, to varying degrees, stemming from linguistic and logical constraints, societal rules, norms, conventions, and their own commitments. Constraints are also material – things act independently of human intentions and do not always cooperate. Material constraints on agency entail not only limits imposed by the things that scientists study, but also the scientist's own physicality. Hence Brinkmann notes that personhood "presupposes a number of things, including linguistic capacities and (of course) a well-functioning brain" (Brinkmann, 2021, p. 23).

Constraints on agency also mean that scientific discovery cannot be merely willed into being. There is a great deal of failure in science, and with it, frustration, discouragement, resentment, and resignation. Moreover, agency does not mean that ideas are all consciously directed:

> The conditions in which discovery usually occurs and the general way of its happening certainly show it in fact to be a process of emergence rather than a feat of operative action … guided not so much by the potentiality of a scientific proposition as by an aspect of nature seeking realization in our minds.
>
> *Polanyi, 1946/1964, pp. 33–35*

Nonconscious processes are often key to this emergence, as is well established in studies of the role of incubation in various forms of insight (e.g., Dorfman, Shames, & Kihlstrom, 1996; Hélie & Sun, 2010; Segal, 2004). Polanyi's term is "tacit;" a large portion of his detailed analysis is devoted to the importance of tacit processes of many kinds in the acts that constitute scientific knowing (Polanyi, 1966, 1974/1958).

Personology and the Study of Lives

"Person" is the root not only of personalism and personism but also of personology, a tradition within personality psychology for which the unit of analysis is not a set of characteristics in isolation but rather "the study of human lives and the factors that influence their course" (Murray, 2008/1938, p. 4). This tradition is grounded in the pioneering theoretical and empirical contributions of Henry Murray (2008/1938) and focused on the uniqueness of each human being when considered in the context of the life as a whole (e.g., see Alexander, 1990; McAdams, 1988; White, 1975). Note that there are other approaches to understanding the interrelation of personality, development, and scientific achievement (e.g., Feist, 2006), but personology represents a unique tradition within the broader subfield of personality psychology. Given the concern with the uniqueness of persons understood in the context of lives, the personological tradition more recently emergent in the narrative tradition (e.g., Josselson & Lieblich, 1993; McAdams, 1988, 2008; McLean, 2017) and psychobiography (e.g., Anderson, 2005; Runyan, 1982; Schultz, 2005). Both, in turn, overlap in important ways with earlier psychoanalytic studies of historical lives (e.g., Erikson, 1958, 1969; Freud, 1947). They analyze life as experienced, but also patterns of behavior and response that may escape the person's awareness, all in the context of the social and cultural world and matrix of relationships within which the uniqueness is made manifest over time.

The implications of psychobiography for the psychology of science are explored most recently by Runyan (2013), who though cautious about drawing generalizations about science in general from the study of individual lives, nevertheless, recognizes that "understanding relations between life and work can help in understanding the sources and meanings of a theory" (Runyan, 2013, p. 353). He qualifies this position, however, by acknowledging his conviction that "personal experience can be a source of great insights, or of great errors, and that identifying personal, social, or cultural sources of a theory does not answer question about its more general validity" (Runyan, 2013, pp. 353–354). Such a view accords with our own position that science cannot be reduced to psychological factors; our aim is not to validate theories but merely to illustrate that psychological processes in various categories are always involved in the work of science, including those practices that may enable validation.

Method and Limitations

The chapters in this book (with one exception) do not profess to meet the conceptual or methodological demands of psychobiography. It is safest to say that they are inspired by the holistic aims of personology. Each chapter brings a psychological lens to science by focusing on the interrelations of life and work for a particular scientist, an acting person. Though each has a different emphasis, they all aim for descriptions that incorporate not only personal, social, and cultural, but also cognitive dimensions of scientific practice as evident in the context of each life.

The chapters are framed as case studies in a broad sense, with the unit of analysis being persons participating in science. They may also be called case studies of the psychological dimension of science. Because "the person" and "the personal" are central to what we conceive as the psychological dimension of science, the distinction makes little difference here. According to Yin, a case study "investigates a contemporary phenomenon in depth and within its real-life context, especially when the boundaries between phenomenon and context are not clearly evident" (Yin, 2009, p. 18). The studies in this book concern scientists of the past, but their goal is to show that boundaries between the phenomenon (the set of activities that constitute science) and the context in which the science was conducted are blurred. Moreover, they are not "histories" per se, because the purpose is to analyze specific phenomena within the context of the life of the scientist. The studies draw from biographical material but are thematically organized; they are not new biographical contributions. Although tasked in principle with analyzing the full psychological context of the scientist's work, authors are obviously able to present only a snapshot of that context, some carefully selected details of the scientist's life and work. Following Murray, we recognize that "[i]t is not possible to study all human beings or all experiences of one human being. The best that can be done is to select representatives or specially significant events for analysis and interpretation" (Murray, 2008/1938, p. 3).

Yin (2009, 2012) distinguishes kinds of case study based on the purposes to which they are applied. Because the case studies in this book do not develop theory, they are most accurately classified as descriptive, focusing on "what has happened" (given the focus on scientists of the past) rather than how or why it happened (Yin, 2012). This is not to say that they do not draw out theoretical implications, but they do not make causal claims about the influence of psychological processes on scientific discovery or theory validation. Moreover, the studies in this volume are presented primarily "for discussion and debate among students," for which reason they might best be regarded as examples of the "teaching case" (Yin, 2009, p. 5) rather than the use of case study as a rigorous research method. Nevertheless, the chapters reflect the "researcher as instrument" stance that recognizes the author of each case study as an agent making decisions about what to include and how to present it from the sources available. In keeping with this stance, each study should be taken as a single point of view on the life and work of the scientist described. Importantly, "person-centered" studies consider not only the personhood of the subject matter, but of the interpreter as well.

The projects bear enough in common with psychobiography, however, to be subject to similar logical, interpretive, and ethical concerns (see Ponterotto, 2015, 2017; Rosenwald, 2012). One concern is with the question of accuracy of portrayal. Although the authors embrace their selective, interpretive position in relation to their subjects, the lives they study are real, and the interpreter risks presenting factual material about lives in an inaccurate way. Moreover, the organization and the emphasis of biographical details may not reflect the experience of the person depicted. Similarly, the studies are not intended as personality assessment, even if they include personality attributions provided by biographers or other sources. Because they are describing "the personal" aspects of the scientist's life, and because they concern "historical actors," they are subject to the challenges Craik describes in relation to "assessing the personality of historical figures," and the "age old problem of how best to describe a person" (Craik,1988, p. 196). This is, of course, an issue for any biographical study, and Craik notes Nicholson's view of the "goal of balancing empathy and esteem with realism" in that genre (Craik, 1988, p. 200). Craik presents several procedures for conducting a systematic and comparative analysis of the personality of historical actors. Chapter authors walk a similar line between sympathetic engagement and critical, distanced evaluation with the scientist described. Moreover, because authors chose the case studies based on their own interest or sense of resonance with historical scientists, the balance is on the side of sympathy and esteem in most of the work here. Nevertheless, the intent is not to offer a "great man" history in accounting for scientific achievement but rather, as noted, to provide illustrations of psychological processes that infiltrate and surround it, with the psychological to be understood as inclusive of personal, social, and cognitive dimensions.

An additional challenge facing the psychologist attempting to interpret the life of a historical scientist is limited knowledge about the scientist's field of practice. History and philosophy of science are aided by educational training not just in history and philosophy but also by at least some mastery of the science analyzed. In the case of this project, the authors do not have degrees in the sciences described and, in some instances, this has posed difficulties in their efforts to bring out the psychological dimensions of the work. On the positive side, given that the purpose of these studies is principally educational, they also provide an excellent path to understanding more about the science represented, lending an interdisciplinary benefit to the person-centered case study.

Sources in the case studies present another obvious limitation. The students are not trained historians and do not have access to archival materials. In some cases, the available biographical materials were very scare, and the student was reliant on one or two primary texts from which to draw relevant biographical details. Authors made note of these limitations up front and acknowledged their indebtedness to the authors on which they relied.

The case studies do contribute examples of ways students can engage science psychologically, and through channels other than presentation of research methods for social science. They offer ways to think about how cognitive, social, and personal factors are integrated in science practice, and how they might be explored more holistically in the context of the study of lives. They are intended to spark reflection and discussion and may serve as examples for students to embark on their own person-centered studies of science, historical or contemporary, and including psychological scientists. Nevertheless, given the limitations of this project and of interpreting historical figures more generally, "modesty is clearly the order of the day" (Craik, 1988, p. 197).

Summary of Contributions

With one exception, the studies in this volume are organized in chronological order (by birthdate) to invite reflection on changes in the activities of scientists over time. However, chapters also take up different themes, depending on the author's interest and emphasis. Each begins with a "personal preamble" describing the author's reason for selecting the scientist discussed in that chapter, including relevance to the author's research projects, felt resonance or connection with the scientist, or assumptions to be drawn from the scientist's work that accord with the author's assumptions about the nature of knowledge. In this sense, the personal preambles are offered in the interests of the reflexivity important to any interpretive work (e.g., see Levitt, 2020). Each case study is also followed by several discussion questions.

The first chapter examines German astronomer *Johannes Kepler* (1571–1630), and calls into question received views about the incompatibility of spirituality with logic and scientific rigor. Azam Khalid highlights Kepler's deep religious commitments, which he shows to be infused in Kepler's dedication to exploration of the universe. Khalid notes that the philosopher he most admires, Charles Sanders Peirce, considers Kepler an exemplary scientific reasoner, and Khalid probes the complex connections between Kepler's faith and reason with this in mind. In so doing, Khalid helps to clarify the nature of abductive reasoning and the role of habit and relates these concepts to some of Kepler's important discoveries and theoretical advances.

Michael Steder then explores the life and work of *Johann Wolfgang von Goethe* (1749–1832), who, though best known for his poetry and literature, also made important contributions to science, especially through a novel theory of optics. Steder foregrounds Goethe's engaged, passionate relation to his subject matter and his lifelong devotion to principles of harmony and balance. As Steder brings out, these imbue Goethe's scientific method with an aesthetic quest that ultimately enhances his ability to make careful and detailed observations of the natural world, from which he draws generalities.

The important role of aesthetic preferences in scientific reasoning is also a prevalent theme in Ahmed Asad's discussion of Russian chemist *Dimitri Mendeleev* (1834–1907). Asad additionally emphasizes the vital role of Mendeleev's historical and local cultural context, along with his values, cognitive propensities, and personal dispositional qualities in specific episodes of his most effective reasoning. In addition, Asad emphasizes the essential semiotic dimension of science through analysis of Mendeleev's engagement with representations, relating the periodic table to Peirce's concept of the icon. Also of interest is Asad's attention to a dream Mendeleev had on the day of his most important discovery. Although the relation of this dream to the discovery itself is a source of some controversy, Asad draws from Henri Poincaré to suggest a cooperative operation of conscious goal-directed and unconscious processes in Mendeleev's eventual insight.

Henri Poincaré (1854–1912) is himself the subject of Yousaf Raza's chapter. Best known as a mathematician, Poincaré also made important contributions to physics. In addition to describing his life and the context it provides for understanding Poincaré's contributions, Raza examines Poincaré's own analysis of his creative process, for which reason he can be viewed as making an important contribution to the psychology of science in his own right. Raza explores Poincaré's views on the nature of mathematical creativity, probing the part played by what the mathematician considered to be intuition. Raza relates Poincaré's view that the most useful of possible combinations of ideas ultimately cross the threshold of consciousness to important insights from pragmatist theory. He argues that the epistemological implications of Poincaré's insights have application for understanding intuition in mathematical reasoning more broadly.

The exception to the chronological order of chapters is in following discussion of Poincaré with a chapter on *John Forbes Nash Jr.* (1928–2015), the pioneering complex systems theorist and mathematician. Ronald Hopkin's chapter on Nash similarly discusses mathematical intuition, so it is useful to consider it in comparison with the chapter on Poincaré. Hopkins probes the nature of Nash's monumental insights and the competing influence of psychopathology on his reasoning. Hopkins, unlike the other chapter authors, deliberately frames his analysis of Nash as psychobiography, drawing upon a developmental approach he finds appropriate for the purposes. Hopkins explores the loneliness of Nash's mathematical genius and highlights Nash's troubled relationship with his *own* consciousness. The chapter examines strategies Nash employed to combat intrusive thoughts and the supportive role of his family in helping him to continue to hone his extraordinary insights despite extraordinary personal challenges.

Georgia Crowe's discussion of *Franz Boas* (1858–1942) introduces several themes relevant to the remainder of the book, including the idea that science is valuable not only for discovery but for application, to address problems in the world and to make things better. Crowe describes not only Boas' pivotal influence on shaping the contemporary field of anthropology, she also brings out his use of systematic method to counter the racist hereditarian assumptions disguised as science in his time. She notes how Boas, through empirical studies, profoundly influenced our contemporary understandings of race as a social phenomenon. Crowe makes an important connection between Boas' scientific activism and feminist epistemology, highlighting the perspectival nature of knowing and challenging taken-for-granted assumptions about objectivity and the stance of the scientist in relation to the subject of study. Finally, the chapter invites comparison of natural and social science.

Scientific activism and the engaged relation with the subject matter of the scientist also characterize the last two chapters. The chapters feature the contributions of the only two female scientists described in this book.

Peder Schillemat looks at the life of marine biologist and conservationist *Rachel Carson* (1907–1964). In addition to examining early experiences and social relationships that influenced Carson's career trajectory, Schillemat comments on the challenges and discrimination Carson faced as a woman in science at the time. She also faced severe political backlash for merely publishing findings that demonstrated environmental damage caused by industry practices. Given the popular attention to Carson's writings, Schillemat questions the forms through which science is most effectively communicated. Carson's passionate dedication to conservation throughout her career blurs boundaries between science and advocacy. The chapter raises questions about relations between science and policy, science and values, and more broadly, about the distinction between science and everyday life.

Dian Fossey (1932–1985) is the subject of Stephen Antczak's chapter in our final case study. Antczak foregrounds the intimate relation between Fossey and her subject matter (gorillas) in an especially vivid challenge to the ideal of scientific detachment. At the same time, he points to the careful observational procedures she used as a basis for drawing generalizations and making predictions about the outcome of poaching and other harmful practices, and this despite the lack of formal training as a scientist. Antczak also acknowledges the fruit of Fossey's labor on behalf of gorillas and the deep interconnections of her work as both a researcher and conservationist. In developing these themes, Antczak acknowledges social relationships that supported or threatened Fossey's work as a scientist/activist. Importantly, he also comments on the sources of funding (in part made possible through relationships) that sustained her ongoing efforts to preserve and protect her subjects.

Following the case studies, Professor Ronald Miller offers a chapter of commentary. Miller, a clinical psychologist with philosophical training, is well known for his thoughtful and rigorous contributions to the use of case study method in psychology. The commentary identifies three themes that emerge across the author's reading of the case studies, helping to clarify and extend their implications. Miller's themes and his reflection on the moral dimension of science help to connect the case studies more cohesively and extend the implications of the project in very contemporary directions.

Final Comments

The case studies present a wide range of scientists, from different disciplines, at different times, and with different subject matter. Readers are invited to reflect on important similarities and differences across these examples, and to more broadly consider how life, science, theory, and policy intermingle in every case.

We wish we could include more chapters on female scientists. This is a problem we readily acknowledge and regret. Similarly, only one chapter is authored by a woman (Georgia Crowe), though she makes the most of this position by deftly weaving feminist epistemology into her study of a male social scientist.

We also regret the lack of inclusion of scientists of color and indigenous science. Three of the chapters are written by authors originally from Pakistan. Their choice of scientists is "Western," as is their philosophical alignment with pragmatism, but undoubtedly their values and commitments reflect their own cultural grounding and, we think, enhance their creative engagement with their subject matter.

A brief reflection on the overall message we wish to convey with these case studies is in order as we close this introduction. As noted, we are not in the business of developing theory from them, but it is important to be transparent about our own assumptions regarding our subject matter.

The admission that science is a deeply human activity, as we hope each chapter makes clear, may tempt the conclusion that science is therefore *no* different from any other activity, that it is just another point of view, on equal par with opinion, no more trustworthy than bare assertion and hearsay. That is not what we wish to convey, and in fact, we consider this a dangerous conclusion. Acknowledging that science involves imagination, creativity, private experience, social negotiation, and the mark of a culture, does not imply that it is reducible to any of these things. Nor does the contemporary view that science consists in fallible theories, always open to modification or abandonment, mean that it lacks any epistemic authority. Its messy humanness is precisely the reason scientific theories must continually undergo various tests, checks, extensions, and refinements. Science offers traditions, tools, and instruments to approach, at least, if not uncover, the secret workings of nature and to address human problems great and small. It does so by means of innately human capacities for making observations and reasoning about them, experiencing flashes of inspiration, and communicating insights to others. It does so within a socially developed system of checks and balances that, on a good day, give us reason for at least a tentative trust in the better grounded of its conclusions. Importantly, though, science, like other human activity, may be directed toward good or ill, the nature of which requires continual reflection and debate. In a broad sense, then, it may be considered a moral activity, or an activity with profound moral dimensions. If psychology concerns itself with science (as we think it should), this moral dimension is always at least implicit.

We end with this summary from Henry Cowles, offered in the context of an excellent recent historical review of conceptions of scientific method: he notes that we might "think of science as the flawed, fallible activity of some imperfect, evolving creatures *and* as a worthy, even noble pursuit" (Cowles, 2020, p. 279, emphasis added).

References

Alexander, I. E. (1990). *Personology: Method and content in personality assessment and psychobiography*. Duke University Press.

Allport, G. W. (1947). Scientific models and human morals. *Psychological Review, 54*(4), 182–192. https://doi.org/10.1037/h0059200

Anderson, J. W. (2005). The psychobiographical study of psychologists. In W. T. Schultz (Ed.), *Handbook of psychobiography* (pp. 203–209). Oxford University Press.

Bergman, L. R., & Andersson, H. (2010). The person and the variable in developmental psychology. *Zeitschrift für Psychologie/Journal of Psychology, 218*, 155–165.

Bickhard, M. H. (2017). How to operationalize a person. *New Ideas in Psychology, 44*, 2–6.

Brinkmann, S. (2021). Minds, brains, or persons? What is psychology about? In B. D. Slife, F. C. Richardson, & S. C. Yanchar (Eds.), *Routledge international handbook of theoretical and philosophical psychology: Critiques, problems, and alternatives to psychological ideas* (pp. 13–29). Routledge.

Clegg, J. W. (2022). *Good science: Psychological inquiry as everyday moral practice*. Cambridge University Press.

Cowles, H. (2020). *The scientific method: An evolution of thinking from Darwin to Dewey*. Harvard University Press. https://doi.org/10.4159/9780674246843

Craik, K. H. (1988). Assessing the personalities of historical figures. In W. M. Runyan (Ed.), *Psychology & historical interpretation* (pp. 196–218). Oxford University Press.

Dorfman, J., Shames, V. A., & Kihlstrom, J. F. (1996). Intuition, incubation, and insight: Implicit cognition in problem solving. In G. D. M. Underwood (Ed.), *Implicit cognition* (pp. 257–296). Oxford University Press.

Dunbar, K., & Blanchette, I. (2001). The in vivo/in vitro approach to cognition: The case of analogy. *Trends in Cognitive Sciences, 5*(8), 334–339.

Eiduson, B. T. (1962). *Scientists: Their psychological world*. Basic Books.

Erikson, E. (1958). *Young man Luther*. W. W. Norton & Co.

Feist, G. J. (1998). A meta-analysis of personality in scientific and artistic creativity. *Personality and Social Psychology Review, 2*(4), 290–309.

Feist, G. J. (2006). How development and personality influence scientific thought, interest, and achievement. *Review of General Psychology, 10*(2), 163–182. https://doi.org/10.1037/1089-2680.10.2.163

Feist, G. J. (2008). *The psychology of science and the origins of the scientific mind.* Yale University Press. https://doi.org/10.12987/9780300133486

Feist, G. J., & Gorman, M. E. (2013). Introduction: Another brick in the wall. In G. J. Feist & M. E. Gorman (Eds.), *Handbook of the psychology of science* (pp. 3–19). Springer Publishing Company.

Freud, S. (1947). *Leonardo da Vinci: A study in psychosexuality.* Crown Publishing Group/ Random House.

Gopnik, A. (2012). Scientific thinking in young children: Theoretical advances, empirical research, and policy implications. *Science, 337*(6102), 1623–1627.

Harré, R. (1984). *Personal being: A theory for individual psychology.* Harvard University Press.

Hélie, S., & Sun, R. (2010). Incubation, insight, and creative problem solving: A unified theory and a connectionist model. *Psychological Review, 117*(3), 994–1024. https://doi. org/10.1037/a0019532

James, W. (1975). *Pragmatism.* Harvard University Press. Originally published 1907.

Josselson, R., & Lieblich, A. (Eds.). (1993). *The narrative study of lives.* Sage Publications.

Klahr, D., & Dunbar, K. (1988). Dual space search during scientific reasoning. *Cognitive Science, 12*(1), 1–48.

Klahr, D., & Simon, H. A. (1999). Studies of scientific discovery: Complementary approaches and convergent findings. *Psychological Bulletin, 125*(5), 524–543. https://doi. org/10.1037/0033-2909.125.5.524

Lamiell, J. T. (2012). Introducing William Stern (1871–1938). *History of Psychology, 15*(4), 379–384. https://doi.org/10.1037/a0027439

Lamiell, J. T. (2021). *Uncovering critical personalism: Readings from William Stern's contributions to scientific psychology.* Springer Nature.

Levitt, H. M. (2020). *Reporting qualitative research in psychology: How to meet APA style journal article reporting standards.* American Psychological Association.

Lundh, L. G. (2015). The person as a focus for research—The contributions of Windelband, Stern, Allport, Lamiell, and Magnusson. *Journal of Person-Oriented Research, 1*(1–2), 15–33.

Mahoney, M. J. (1976). *Scientist as subject: The psychological imperative.* Ballinger.

Mahoney, M. J. (1979). Review paper: Psychology of the scientist: An evaluative review. *Social Studies of Science, 9*(3), 349–375. https://doi.org/10.1177/030631277900900304

Martin, J. (2021). A non-reductive "person-based ontology" for psychological inquiry. In B. D. Slife, F. C. Richardson, & S. C. Yanchar (Eds.), *Routledge international handbook of theoretical and philosophical psychology: Critiques, problems, and alternatives to psychological ideas* (pp. 391–411). Routledge.

Maslow, A. (1966). *The psychology of science: A reconnaissance.* Harper & Row.

McAdams, D. P. (1988). *Power, intimacy, and the life story: Personological inquiries into identity.* Guilford press.

McAdams, D. P. (2008). Personal narratives and the life story. In O. P. John, R. W. Robins, & L. A. Pervin (Eds.), *Handbook of personality: Theory and research* (pp. 242–262). The Guilford Press.

McLean, K. C. (2017). And the story evolves: The development of personal narratives and narrative identity. In J. Specht (Ed.), *Personality development across the lifespan* (pp. 325–338). Elsevier.

Miller, R. B. (2004). *Facing human suffering: Psychology and psychotherapy as moral engagement.* American Psychological Association.

Miller, R. B. (2011). Real clinical trials (RCT')—Panels of psychological inquiry for transforming anecdotal data into clinical facts and validated judgments: Introduction to a pilot test with the case of "Anna". *Pragmatic Case Studies in Psychotherapy, 7*(1), 6–36.

Mitroff, I. (1974). *The subjective side of science: Philosophical inquiry into the psychology of the Apollo Moon scientists.* Elsevier.

Murray, H. (2008). *Explorations in personality* (70th anniversary edition). Oxford University Press. Originally published 1938.

Nersessian, N. J. (1984). *Faraday to Einstein: Constructing meaning in scientific theories.* Martinus Nijhoff/Kluer.

Nersessian, N. J. (2022). *Interdisciplinarity in the making: Models and methods in frontier science.* MIT Press.

O'Doherty, K. C., Osbeck, L. M., Schraube, E., & Yen, J. (2019). Introduction: Psychological studies of science and technology. In K. C. O'Doherty, L. M. Osbeck, E. Schraube, & J. Yen (Eds.), *Psychological studies of science and technology* (pp. 1–28). Palgrave Macmillan.

Osbeck, L. M., & Nersessian, N. J. (2017). Epistemic identities in interdisciplinary science. *Perspectives on Science, 25*(2), 226–260.

Polanyi, M. (1964). *Science, faith, and society.* University of Chicago Press. Originally published 1946.

Polanyi, M. (1966). *The tacit dimension.* University of Chicago Press.

Polanyi, M. (1974). *Personal knowledge: Toward a post-critical philosophy.* University of Chicago Press. Originally published 1958.

Ponterotto, J. G. (2015). Psychobiography in psychology: Past, present, and future. *Journal of Psychology in Africa, 25*(5), 379–389.

Ponterotto, J. G., & Reynolds (Taewon Choi), J. D. (2017). Ethical and legal considerations in psychobiography. *American Psychologist, 72*(5), 446–458. https://doi.org/10.1037/amp0000047

Reber, J. S., & Slife, B. D. (2021). Psychology's flawed focus on individuals and individualism: A strong relationality alternative. In B. D. Slife, F. C. Richardson, & S. C. Yanchar (Eds). *Routledge international handbook of theoretical and philosophical psychology* (pp. 30–54). Routledge.

Rogers, C. R. (1955). Persons or science? A philosophical question. *American Psychologist, 10*(7), 267–278. https://doi.org/10.1037/h0040999

Rogers, C. R. (1961). *On becoming a person: A therapist's view of psychotherapy.* Constable.

Rosenwald, G. (2012) The psychobiographer's authority: Questions of interpretive scope and logic. *The Psychoanalytic Quarterly, 81*(2), 357–400.

Runyan, W. M. (1982). *Life histories and psychobiography: Explorations in theory and method.* Oxford University Press.

Runyan, W. M. (2005). Evolving conceptions of psychobiography and the study of lives. In W. T. Schultz (Ed.), *Handbook of psychobiography* (pp. 19–41). Oxford University Press.

Schmid, P. F. (1998). On becoming a person-centred approach: A person-centred understanding of the person. In B. Thorne & E. Landers (Eds.), *Person-centred therapy: A European perspective* (pp. 38–52). Sage.

Schultz, W. T. (Ed.). (2005). *Handbook of psychobiography.* Oxford University Press.

Schunn, C. D., & Anderson, J. R. (1999). The generality/specificity of expertise in scientific reasoning. *Cognitive Science, 23*(3), 337–370.

Segal, E. (2004). Incubation in insight problem solving. *Creativity Research Journal, 16*(1), 141–148.

Simonton, D. K. (2004). *Creativity in science: Chance, logic, genius, and zeitgeist*. Cambridge University Press.

Smythe, W. (1998). *Toward a psychology of persons*. Erlbaum.

Sprague, E. (1999). *Persons and their minds: A philosophical investigation*. Routledge.

Tweney, R. Stopping time: Faraday and the scientific creation of perceptual order. *Physis: Revista Internazionale di Storia Dell Scienza, 29*, 149–164.

White, R. W. (1975). *Lives in progress: A study of the natural growth of personality* (3rd ed.). Holt, Rinehart & Winston.

Yin, R. K. (2009). *Case study research: Design and methods* (4th ed.). Sage.

Yin, R. K. (2012). *Applications of case study research*. Sage.

Case Studies

Case Studies

1

JOHANNES KEPLER

A Pragmaticist Priest of God at the Book of Nature

Muhammad Azam Khalid

Personal Preamble

During the Psychiatry residency, one of my favorite past times that I developed, along with two friends of mine, was to tease others about how their diagnoses might be wrong and how some other diagnosis is also possible with the same data. Later, when I started reading Charles Peirce, his theory of inquiry and understanding of the scientific method attracted me the most. In Peirce's term, the joking about diagnoses was to point out how coming to a hypothesis is not a deductive process but an abductive one and that the abductive reasoning is not about security but fecundity of the inquiry. Peirce knew too much about the history of science and was very critical, personally and philosophically. I chose Johannes Kepler as the case study because of Peirce's comments. My curiosity about Kepler can be summed up by the thought that if someone like Peirce judges Kepler as the most remarkable example of abductive reasoning ever, how exceptional a reasoner would he be? Second, Peirce attributed Kepler's success to a moral quality. This second comment does not fit with the typical persona of the scientist. The moral qualities typically thought of as a scientist pertain to her relationship with peers and society. How can personal commitment to moral or any other extra-scientific principles make someone the greatest abductive reasoner of all times? This question guides my inquiry into Kepler's life and the psychological dimensions of his accomplishments. To state it differently, how does personal affect the allegedly impersonal, rational aspects of scientific inquiry?

DOI: 10.4324/9781003276692-3

Introduction

Johannes Kepler's influence on modern science is far more significant than Copernicus or Galileo, yet, apart from his three laws, he, as a person, is little known even in academic circles. It was Kepler who paved the way for Newton's work. For any science student, the story of modern scientific astronomy started with Copernicus. However, in his *De Revolutionibus Orbium Coelestium* (DR, 1543), Copernicus was actually trying to purge astronomy of Ptolemaic influence by returning to the principles of the ancients rather than moving forward in any way. Thomas Kuhn explains that "in every respect except the earth's motion the DR seems more closely akin to the works of ancient and medieval astronomers and cosmologists" (Kuhn, 1957, p. 134). By contrast in his *Astronomia nova*, or *New Astronomy* (NA, 1609), Kepler wanted to explain the motions of Heavens using mechanical principles of corporeal physics and "mathematically expressed harmonies which man can discover in the chaos of events" (Holton, 1988, p. 78). As I will argue, not only this interest proceeds from personal, religious commitments for Kepler, the analytical tools at his disposal were also rooted in the same commitments. He was first a priest and then an astronomer. Living in a chaotic or incoherent universe, devoid of all order, was not possible for him. Nor was he satisfied with the explanation of souls, for it becomes inconsistent in the face of variability and erratic behavior shown by planets. At the grandest astronomical scale possible in his times, his whole endeavor was an attempt at sense-making directed at the chaos in the universe and rectifying the incoherence created by the theories of astronomy and cosmography of his times. For him, he was fulfilling his duty by answering "the divine voice that enjoins humans to study astronomy" (Kepler, 1609/1992, p. 183)

Due to his interest in the logic of science, Charles Peirce referred to Kepler all his life, primarily to understand and explicate the logical inference he called "abduction" – the logic of formulating a new hypothesis. Not only was he "the greatest reasoner who ever lived" (Peirce, 1958), for Peirce, Kepler was also "one of the most interesting personalities that ever lived." He revered the creator of *New Astronomy* that despite personal deficiencies of bad eyesight and crippled hands due to smallpox, Kepler's work was even more "cunninger than any deciphering of hieroglyphics or of cuneiform inscriptions" (p. 250). Peirce regarded Kepler's work as the best example of abductive inference. Despite this emphasis, there has been no detailed analysis of Kepler's work done by Peirce scholars, except in a dissertation chapter (Silva, 2007). Abductive hypotheses, for Peirce, are generated when our earlier hypotheses are not producing desired effects, that is, explanations.

The purpose of this study into Kepler's work is threefold. First, it is to understand why Kepler is the most remarkable example of reasoning, especially abduction. Second, it is to ask how can Kepler's work help us better understand abduction? Third, and most importantly, it is to explore the psychological and personal dimension of the first part of an inquiry, that is, abduction as enacted by the particular person Johannes Kepler.

Life of Johannes Kepler

Johannes Kepler's life can be presented as an example of almost failures that turned out to be brilliant successes. Not only do different events of his life follow this pattern, but his life, in general, could have been a failure right from the start. Born prematurely, he struggled with his health in childhood (Caspar, 2012) during which he developed nearsightedness and "crippled fingers" due to smallpox, making him "awkward" (Peirce, 1958, p. 251) at taking astronomical measurements. And, yet he changed astronomy forever.

His grandfather, Sebald Kepler, was the city's Lord mayor, but the fortunes and stature of the family dwindled afterward. Johannes' father, Heinrich Kepler, led a life as a mercenary and left the family when Johannes was five years of age; he never returned. Kepler's earliest introductions to astronomy were the sights of the Great Comet of 1577 and the lunar eclipse of 1580 when Kepler was six and nine, respectively (Koestler, 2017). After attending The Protestant Seminary of Maulbronn, Kepler got into the University of Tübingen and studied mathematics and astronomy under Michael Maestlin. Maestlin was one of the first major astronomers who accepted the Copernican heliocentric models (Kuhn, 1957). Since his student days under Maestlin, Kepler adopted the heliocentric model of the universe. At 22, in 1594, Kepler was recommended, against his wish, for a position to teach mathematics and astronomy at the Protestant school of Graz (Caspar, 2012), a move Kepler attributed to the divine providence (Kepler, 1609/ 1992), as it changed his life in significant ways. In Graz, he wrote his first major work, *Mysterium Cosmographicum* (*Mystery of Universe*, MC), in 1596, at the age of 24, which led his introduction to the greatest astronomer of the age Tycho Brahe. He also got married, in Graz, to Barbara Muller in 1595. Barbara was a 23-year-old, twice-divorced daughter of a mill owner who brought a good deal of fortune with her, along with a daughter. Kepler's introduction and a subsequent visit to Tycho Brahe got him a position in Prague as an assistant to Brahe. This move brought financial stability and religious relief to Lutheran Kepler's life as it was the period of counterreformation sentiment in Graz.

Although the Thirty Years' War had not started yet, there was tension in the air. Had Kepler converted to Catholicism, which he was given a choice to do, he would not have had to leave Graz. However, Prague was not as affected by the tension, at least not as much as Graz was. Brahe kept his carefully gathered data on the positions and movements of the heavens very close to his chest and did not let anyone see it. Those measurements ensured Brahe's position in the court for a long time. After the death of Brahe in 1601, Kepler was promoted to his place at the court as the Imperial Mathematician of the Holy Roman Empire (Caspar, 2012). Brahe willed that his notebooks having the data on Mars must go to his family, but Kepler stole the notebooks to test his theories (Teller, Teller, & Talley, 1991). In the next four years, Kepler resolutely worked to win his "war on Mars," discovering its orbit and the physics of the heavens (Caspar, 2012, p. 133). Still, he could not publish his *Astronomia Nova* until 1609 due to the dispute with Brahe's family over Kepler's use of Brahe's measurements.

Kepler discovered his first two laws and provided their proofs in the *New Astronomy*, 1609, but the third law was not discovered until 1619 in *Harmonices Mundi* (*Harmonies of the World*). The 15 years from 1595 to 1610 brought the best out of Kepler, both personally and scientifically. He had three children with Barbara during these years and had the chance to test his hypotheses formulated in *Mysterium Cosmographicum*, using Brahe's data.

In 1611, Barbara, Kepler's wife, and a son, died from Hungarian spotted fever. Kepler could not keep his position of Imperial Mathematician due to Emperor Rudolf stepping down in favor of his brother Matthias. The Imperial Mathematician was helpless and sought after a job as the district mathematician at Linz (Caspar, 2012, p. 204). He was reinstated as the Imperial Mathematician by Matthias after ascending to the throne of The Holy Roman Emperor. After seeing eleven proposals, Kepler married again in 1613 to Susanna Reuttinger. Three children from this marriage survived to adulthood. Even in choosing his future wife, he was at his creative best. His method to choose among the marriage proposals was used by Ferguson (1989) to solve the "secretary problem." Besides astronomy, Kepler also suggested innovations in optics to make telescopes more powerful and formulated the Kepler Conjecture. Johannes Kepler died on 15 November 1630. His grave was destroyed in the Thirty Years War, and only a self-authored epitaph survived:

"I measured the skies, now the shadows I measure
Skybound was the mind, earthbound the body rests"

(as cited in Koestler, 2017, p. 422)

Kepler's Religiosity

Kepler was a devoutly religious person. He never wanted to be a person of science but a Lutheran minister. His writings are filled with religious references. When he was nominated against his wish for the position in Graz, astronomy became a theology for him: "What voice have the stars, to praise God as man does? Unless, when they supply man with the cause to Praise God, they are said to praise God" (Kepler, 1596/1981, p. 53). Just as a theologian approaches Scripture to interpret God's mind expressed as words, Kepler sat in front of Nature, "as a priest of God at the Book of Nature" (as cited in Kozhamthadam, 1994, p. 41), to interpret God's mind expressed as dynamics of heavens. Kepler understood his calling toward astronomy as a divine decree to decipher the secrets of heavens to call people toward the most wise creator. In his introduction to the *New Astronomy* (1609/ 1992), he invites his reader to "praise and celebrate the Creator's wisdom and greatness" that he, just like a priest, is going to unfold for him in the more perspicacious explanation of the world's form, the investigation of causes, and the detection of errors of vision. Let him not only extol the Creator's divine beneficence in His concern for the well-being of all living things, expressed in the firmness and stability of the earth, but also acknowledge His wisdom expressed in its motion, at once so well hidden and so admirable (Kepler, 1609/1992, p. 65).

Kepler's scientific achievements can be summarized as inquiries into the hypotheses proposed in *Mysterium Cosmographicum* when he was still in Graz, without Brahe's evidence. Furthermore, he had no reluctance to admit, "nearly all the books on astronomy which I have published since then were related to one or the other of the main chapters in this little book and are more thorough expositions or completions of it" (as cited in Koestler, 2017, p. 260). He had both physical and metaphysical reasons to write this book. He never abandoned these metaphysical reasons; instead, he argued for them all his life. He was committed to providing proof for the heliocentric universe, and according to Kepler, while "Copernicus did so through mathematical arguments, mine were physical, or rather metaphysical" (Kepler, 1596/1981, p. 62). By metaphysical, he does not mean a nonreligious metaphysics but a reference to "the Father, and the Son, and the Holy Spirit" (p. 63). He did not only pursue this resemblance in this book only but all his life. When Kepler first had the idea of explaining the arrangement and number of planets around Sun on 19 July 1595, while teaching, he felt as if "an oracle had spoken to him from heaven" (Caspar, 2012, pp. 62–64) about the idea that the five platonic solids.

Kepler's investigation of Mars's orbit and not that of any other planet was not a result of careful consideration. It was pure chance. When Kepler arrived in Prague, Brahe was working on Mars, the planet with the most elliptical orbit among planets known at that time:

> I therefore once again think it to have happened by divine arrangement, that I arrived at the same time in which he was intent upon Mars, whose motions provide the only possible access to the hidden secrets of astronomy, without which we would remain forever ignorant of those secrets.
>
> *Kepler, 1609/1992, p. 185*

Every move in his scientific inquiry or life, in general, that could not be explained mathematically or geometrically, he explained religiously. He was thus thematically committed to Lutheran theology in all his sense-making activities: "In the end, Kepler's unifying principle for the world of phenomena is not merely the concept of mechanical forces, but God, expressing Himself in mathematical laws" (Holton, 1988, p. 85).

Kepler's Process of Discovery

Both Kepler's discoveries and the process of reaching those discoveries have been discussed extensively by philosophers and historians of science. His methods to approach the problems and inquiry has been studied as the prototype of scientific inquiry by William Whewell, John Stuart Mill, Charles Sanders Peirce, Karl Popper, Alexander Koyre, Russel Norwood Hanson, Arthur Koestler, Gerald Holton, Curtis Wilson, and others. Moreover, there are as many views about the Keplerian method as there have been students of the method. Newton was the first to suggest that Kepler did not have any method but instead arrived at his hypotheses as lucky guesswork and "did not deserve the credit for their discovery" (Kozhamthadam, 1994, p. 2). Curtis Wilson described Kepler's hypotheses as "hunches" about the dynamics of a phenomenon (Wilson, 1972). For Koestler, Kepler's inquiry is akin to "sleepwalking," like groping for something in the dark and stumbling upon it by chance (Kosetler, 2017). On the other extreme, some have tried to reduce his whole inquiry into an algorithmic procedure directed by Tycho Brahe's data (Whiteside, 1974). Edwin Brutt went as far as calling him a "sun worshipper" (Kozhamthadam, p. 4). Other than Koestler (1961), no one discussed the psychological aspect of Kepler's discovery, but that too is a very crude formulation of such a dynamic process. For Kepler himself, "[T]he roads by which men arrive at their insights into celestial matters seem to me almost as worthy of wonder as those matters in themselves" (cited in Koestler, 2017, p. 261).

Before Kepler, "astronomy" was different from our contemporary use of the term. Astronomy, cosmography, and physics were three distinct disciplines and had no relation. The astronomer was supposed to provide geometrically measurable positions of the stars and the planets and was mainly used as a service to astrology. Strictly speaking, an astronomer was not supposed to speculate about the causes and dynamics of the motions of the heavens. It was the cosmographer's speculation that discussed these issues. In this sense, Ptolemy, Copernicus, and Brahe were the astronomers working within the limits of Platonic cosmography. They were supposed to come up with the most accurate models for the movement of the heavens. They all agreed that these motions were governed by Platonic souls working in the spheres. Contrarily, physics was related to the corporeal objects and not to the sublime heavens, hence the confusion of Kepler's teacher when he wrote, "I simply do not understand this" (Kozhamthadam, 1994, p. 103). On the other hand, Kepler wanted a physical theory for the astronomical phenomena and named his magnum opus *New Astronomy*.

Before going into the details of Kepler's work, it would be apt to discuss a few of his personal commitments. Kozhamthadam (1994) and Caspar (2012) have presented excellent details about Kepler's view of God, the universe, and humans. For Kepler, God is a supremely rational being that does not do anything without "reason." In this reason lies the human opportunity to know the mind and purposes of God. For him, "the Creator chose nothing without a plan" (cited in

Kozhamthadam, p. 19). As we shall see, this personal commitment to a rational God became his guiding principle to argue for the "real sun" in contrast to Copernicus' "mean sun" and his understanding of the gravitational force. For Kepler, it was unreasonable to put the sun in the center if it has no function that is central to all the planets. If for a priest the medium to know God's mind is His words in the form of Scripture, then for the "priest at the book of nature" the medium to know God's actions manifested in the heavens is geometry: "Geometry is co-eternal with the mind of God before the creation of things; it is God himself (what is in God that is not God himself?)" (cited in Kozhamthadam, 1994, p. 20). Geometry, then, is the second personal commitment of Kepler. For him, geometrical principles are the means of knowing God's purposes expressed in the universe. It goes further than that. All his life, Kepler repeatedly brought up "the Sphere," the most perfect solid, as a model to understand the Trinity. The center of the sphere represents God while the surface is the Son and the intervening space of infinite angles the Holy Ghost. Just as God inspires the Son through the Holy Ghost, The Sun in the middle of the moving stars, Himself at rest and yet the source of the motion, carries the image of God, the Father and Creator. He distributes his motive force through a medium which contains the moving bodies, even as the Father creates through the Holy Ghost (as cited in Koestler, 2013, pp. 261–262).

For any planet, it is a Trinity of sun, that planet, and the intervening force. These were the metaphysical reasons for Kepler to argue for a heliocentric universe in the *Mysterium Cosmographicum*. In its introduction, he writes:

> I dared so much was due to the splendid Harmony of those things which are at rest, the sun, the fixed stars and the intermediate space, with God the Father, and the Son, and the Holy Spirit. This resemblance I shall pursue at greater length in my Cosmographia.
>
> *Kepler, 1596/1981, p. 63*

We can see from his own account that what any other "science minded" person might have dismissed as "subjective feelings" or "hunches," Kepler took at face value and pursued in inquiry. Wilson (1972) argues that, "it is an initial hunch, a physical hypothesis, that guides him throughout; every step is taken deliberately, not only in confrontation with the data but also in pursuance of his hunch" (p. 93). How can a scientist do that? More importantly, for our purposes, *why* would he do that? Is this move an example of "sleepwalking" that turned out to be true by chance, or are there any logical warrants for this move? Charles Peirce's understanding of inquiry, especially abduction, and, in it, the central role of a psychological concept, doubt, provides answers to us. For Peirce, "each chief step in science has been a lesson in logic." (CP 5.363), and Kepler's work was, for him, "the greatest piece of Retroductive reasoning ever performed." (CP 1.74)

What does Kepler himself have to say about his "hunches?" He was well aware that many of these hunches would turn out to be wrong. Still, his personal commitment to religion allowed him to consider those hunches worth pursuing and testing: "Not every hunch is wrong. For man is the image of God, and it is quite possible that in regard to certain things that ornament the universe, his opinion is same as God's" (cited in Kozhamthadam, 1994, p. 25). Kepler is arguing for continuity between God's mind and the human mind. It seems that, for Kepler, any movement in his own reasoning about the Heavens was either through mathematical and geometrical tools or by divine providence.

Scientific reasoning belongs to that set of human activities that are "active organismic response to novelty and complexity" (Osbeck, 2014, p. 34). It is an act of sense-making in this regard. Sense-making, in turn, is "organizing, goal-directed accomplishments of embodied and contextually embedded human agents" (Osbeck, 2014, p. 34). The aim of scientific reasoning is to make sense, to rationalize. It is one of many human activities aimed at making sense of the apparent chaos. But what is the immediate motive of this, or any, sense-making activity? In answer to this question lies the answer of why would scientists pursue their hunches? In all his works Kepler used the words "guess," "suspicion," "hunch," "suspect," and "desire" for his hypotheses, yet pursued them seriously. Early in the *New Astronomy* Kepler drew parallels among his work and the likes of Columbus and Magellen (Kepler, 1609/1992, p. 78).

Unlike modern scientists who are "trained on the ascetic standards of presentation originating in Euclid," Kepler, "with rich imagination…frequently finds analogies from every phase of life, exalted or commonplace" (Holton, 1988, pp. 69–70). What warrants does he have to do that? Or in other words, how is this move scientifically and logically justified? Kepler described this motive as the "wonder" for "when experience is seen to teach something different to those who pay careful attention…it gives rise to a powerful sense of wonder, which at length drives men to look into causes" (Kepler, 1609/1992, p. 115). It is wonder that drives an inquirer to look for new explanations: "It is just this from which astronomy arose among men. Astronomy's aim is considered to be to show why the stars' motions appear to be irregular on earth, despite their being exceedingly well ordered in heaven" (p. 115, emphasis added). This description of the purpose of astronomy shows that the aim of Kepler's personal judgment, a hypothesis, a sense-making attempt in itself, was to understand the order in the heavens. It also shows the goal of his attempt at sense-making of the apparent irregular motion of wanderers (literal meaning of the Greek planetae), in the face of earlier hypotheses, that created a doubt in the Kepler's mind, by creating an inconsistency. It is this wonder that an inquiry aims to eliminate by making sense of it either in the light of already held beliefs, changing these beliefs, or discrediting the source of wonder, that is, a novelty in experience. While discussing the Copernican model in *Mysterium Cosmographicum*, Kepler praises Copernicus "for the things at which from others we learn to wonder, only Copernicus…removes the cause of wonder, which is not knowing causes" (Kepler, 1596/1981, p. 75).

Peirce has a similar but much more nuanced understanding of this sense-making process in science. For Peirce, a "real genuine belief, consists in a habit with which one is contented, and which one usually recognizes (though not always) this habit consisting in the general fact that under certain circumstances one would act in a definite way, and would be content to do so" (Robin, 1967, p. 764). Peirce here is talking not only about outward behavior when he mentions "habit" and "act," but also includes habits of experiencing, feeling, thinking, acting, approaching, and positioning. It consists of all activities that could possibly be included. It comprises "any state of mind by virtue of which a person would, under definite circumstances – mostly, if not invariably, consisting in his experiencing conscious experience of some kind – either think, or act, or feel in a definite way" (Robin, 1967, p. 852:10) In that respect, in our case, the methods, preconceptions, philosophical principles, epistemic goals, and rules that an inquirer employ in the inquiry are the results of him having certain beliefs about the phenomenon under investigation. Each belief would continue to produce its effects on the "habit of thought and a conduct" (Peirce, 1976, p. 297) of the inquirer. In this manner, a hypothesis, an attempt at sense-making, is also a belief that, as its desired effects, would enable the inquirer to see the phenomenon in question in a certain way. This hypothesis would position the inquirer to see such patterns in the phenomenon not seeable before and enables her to approach the phenomenon in a novel way in the future. This hypothesis, a belief, would continue to produce its effects accordingly unless encountered by a novel situation that does not allow it to produce the desired effects aimed at sense-making. This was wonder for Kepler, but for Peirce, it is doubt, "A true doubt is […] a doubt which interferes with the proper action of a belief-habit" (Robin, 1967, p. 288:10). Using the word "doubt" makes the relationship between novelty and conduct very explicit, in a way that "wonder" does not. The closest term to wonder that Peirce used is surprise, "The surprising fact, C, is observed; But if A [an explanatory hypothesis] were true, C would be a matter of course, hence, there is reason to suspect that A is true" (Peirce, 1974, 5.189).

Belief and doubt are different in at least three ways for Peirce. Firstly, there is a qualitative difference in subjective feeling: "We generally know when we wish to ask a question and when we wish to pronounce a judgment" (Peirce, 1974, 5.370). Where belief brings a feeling of contentment, doubt brings irritation. Secondly, beliefs are tied to actions. They enable us to develop corresponding habits and conduct while "doubt never has such an effect" (5.371). The third difference is the most important for questions related to us.

Doubt is an uneasy and dissatisfied state from which we struggle to free ourselves and pass into the state of belief, while the latter is a calm and satisfactory state which we do not wish to avoid, or to change to a belief in anything else. On the contrary, we cling tenaciously, not merely to believing, but to believing just what we do believe (5.372).

It is doubt that initiates an inquiry. As far as an inquirer's habits, a sign of her having some beliefs, produces desired results, she remains satisfied with the beliefs. One can see here that Peirce disapproves of armchair doubting as championed by other philosophers, as he is a pragmaticist. When the desired effects are not produced by the habitual conduct, this generates an "irritation of doubt" that forces the inquirer to revisit the beliefs and the resultant habits: "the action of thought is excited by the irritation of doubt, and ceases when belief is attained" (Peirce, 1974, 5.394). Peirce describes four methods of fixing (or refixing) beliefs, that is, tenacity, authority, *a priori*, and the method of science (Peirce, 1877). In their perverse and absolutized (Gilmore, 2003) forms, the first three methods put the inquirer in a very different attitude toward genuine novelty (the situation in which the earlier habits are not producing desired effects) than the method of science. The first question that the practitioner of these three methods would ask would be, how can I doubt the novelty experienced? And then one would go on to reduce the novelty into already held beliefs or explain it way. On the contrary, the attitude of the practitioner of the method of science would be, how can I adapt my beliefs to the novelty experienced? In both cases, the irritation of doubt would be eliminated, in the first instance, by discrediting the novelty in the universe, while in the second, by forming a (new) belief congruent with experience.

The first step of the inquiry, "a struggle to attain a state of belief" (Peirce, 1974, 5.374), initiated by doubt, is abduction. For Peirce, that is "the process of forming an explanatory hypothesis. It is the only logical operation which introduces any new idea; for induction does nothing but determine a value, and deduction merely evolves the necessary consequences of a pure hypothesis" (5.171). Through abduction, an inquirer formulates a new hypothesis. The deductive formulations of the hypothesis are the experiments or possible experiences that the inquirer tests or finds inductively. We cannot go into Peirce's understanding of the Scientific Method here, but some comments about the nature and role of abduction are in order. Abduction aims to provide a new hypothesis that would explain the novelty. Abduction provides us with a possibility of a new belief-habit that would accommodate the novel experience. Abduction merely suggests a possibility and "commits us to nothing (5.602). Just like Kepler's "hunch," that came through an "oracle," the abductive suggestion comes to us like a flash. It is an act of insight, although of extremely fallible insight. It is true that the different elements of the hypothesis were in our minds before; but it is the idea of putting together what we had never before dreamed of putting together which flashes the new suggestion before our contemplation (5.181).

As for Kepler not all hunches are wrong, for Peirce, abductive suggestions are "strong enough not to be overwhelmingly more often wrong than right" in explaining "the general elements, of Nature" (Peirce, 1974, 5.173), There are infinite hypotheses that could explain a phenomenon in any inquiry, but we get to the right one after a dozen false ones at maximum (5.172). One can argue that it is the true belief, not just any belief, that one wants to achieve, but, for Peirce, the concern for truth, for an inquirer in doubt, comes after the concern to attain belief in itself.

We may fancy that this is not enough for us, and that we seek, not merely an opinion, but a true opinion. Put this fancy to the test, however, and it proves groundless; for as soon as a firm belief is reached, we are entirely satisfied, whether the belief be true or false. Furthermore, it is clear that nothing out of the sphere of our knowledge can be our object, for nothing which does not affect the mind can be the motive for mental effort. The most that can be maintained is that we seek for a belief that we shall think to be true. But we think each one of our beliefs to be true, and, indeed, it is mere tautology to say so (5.375, emphasis added).

Among other reasons beyond the scope of this paper, this is why Peirce classifies abduction as an instinct (Peirce, 1974, 5.173) that could be acquired or innate. It is instinct because it is an organismic response to the irritation of doubt that he wants to escape by attaining a state of belief (5.374). Like any other instinct, its first concern is to satisfy its immediate motive, doubt in this case, by fulfilling its aim to employ whatever resources are available at hand. Nevertheless, abduction is not irrational in the traditional sense of the word. It is an "inward power, not sufficient to reach the truth by itself, but yet supplying an essential factor to the influences carrying their minds to the truth" (1.80). There is no way an inquirer can tell if a hypothesis is valid or not before testing its deductive iterations inductively. Only by its effect of explaining the phenomenon can we judge its validity. Its effects on our attitude toward the universe are the measure of its truth. The degree of congruence between the habits of the inquirer and the habits of nature tested inductively is the measure of the confidence we could impart to any hypothesis. It is only later that we could attribute, like Kepler, invalid hunches to our biases: "If I had embarked upon this path a little more thoughtfully, I might have immediately arrived at the Truth of the matter. But since I was blind from desire, I did not pay attention to each and every part" (Kepler, 1609/1992, p. 455, emphasis added).

On the one hand, Kepler argues that "not every hunch is wrong." Still, on the other, he is well aware of the fact that a hunch might be "blind from desire," and just like any other inquirer, only after testing the hypothesis inductively can one make this judgment. This desire is rationalized as one of the three methods of fixing beliefs that Peirce disapproves of. We now know how deeply flawed Ptolemy's model is but for Peirce, his whole inquiry was scientific (Peirce, 1974, 6.428) because, like Kepler, Ptolemy set out to relieve the irritation of doubt produced by the motion of "wanderers," with whatever empirical and philosophical resources he had. The "equant," the "deferent," and the "epicycle" were all abductive attempts aimed to relieve doubt by sense-making of the observed erratic behavior of the planets.

When Kepler started his scientific career, astronomy was still under the influence of Ptolemy. Just a generation earlier, Copernicus presented his model in which earth was not at the center of the universe but had more epicycles than Ptolemy's model, and Tycho Brahe developed a model that is half-Ptolemaic and half-Copernican. Despite all these radical differences, these models were equivalent in the eyes of Kepler. In fact, he named chapter six, part one, of the *New Astronomy*, in which he discussed these models: "On the equivalence of the hypotheses of Ptolemy, Copernicus, and Brahe..." How were they equivalent when we know that Ptolemaic was geocentric and Copernican is popularly regarded as heliocentric? For Kepler, "these three forms are absolutely, perfectly, geometrically equivalent" (Kepler, 1609/1992, p. 157), because for these hypotheses, the heavens must bend to the ideals of Euclidean geometry and Greek perfection, and not the other way around.

The testimony of the ages confirms that the motions of the planets are orbicular. It is an immediate presumption of reason, reflected in experience, that their gyrations are perfect circles. For among figures it is circles, and among bodies the heavens, that are considered the most perfect (Kepler, 1609/1992, p. 115). To explain and make sense of the erratic behavior of planets, these astronomers resorted to adding complexity to their models, but the circle remained there as the ideal because "it was endowed with the authority of all philosophers, and the more convenient it was for metaphysics in particular" (p. 417). In the same chapter, Kepler argues that if adjusted for the "mean sun," all these models can explain the position of Mars at a given place. Still, the problem arises if we stick to any model and follow it. The shape of Mars' orbit would become like a pretzel, and Mars never passes through the same point again (pp. 118–122).

Didn't Kepler have his personal commitments when he started his inquiry? Definitely, he did, as has been discussed. He made it clear that he was still pursuing the hypotheses, let alone commitments, formulated in the *Mysterium Cosmographicum*:

> I entered upon the work girded with the preconceived opinions expressed in my *Mysterium Cosmographicum*. At the beginning there was great controversy between us as to whether it were possible to set up another sort of hypothesis which would express to a hair's breadth so many positions of the planet, and whether it were possible for the former hypothesis to be false despite its having accomplished this so far over the entire circuit of the zodiac.
>
> *Kepler, 1609/1992, p. 185*

We will see how his fixation on his hypotheses was not perverse. We can also see an expression of doubt in this statement. If a hypothesis has been explained and predicted so much over the course of centuries, how could it be false? This had certainly made him doubt his "hunches" that maybe he was thinking clearly, but he pursued them nonetheless. In Kepler's method, what made him leave or modify his preconceptions? It was his attitude toward his abductions, his hunches.

Kepler's Abductive Hypotheses

A disjointed way of looking at his distinct, but interlinked hypotheses, is not appropriate to describe Kepler's approach but is being employed for the purposes of intelligibility.

Corporeal Physics for Heavens

As we have seen, in Kepler's times, astronomy and physics were mutually exclusive disciplines. But for Kepler, the most important question for *New Astronomy* is "*A quo movementur planetae?*" (what is it that is moving the planets?) (cited in Kozhamthadam, 1994, p. 182). Earlier explanations appealing to the Holy Ghost, which he formulated himself, and souls were not convincing to him. Kepler wanted to approach the heavens dynamically to understand the causes, and get to the explanations that remove incoherence in the models. He was hypothesizing a continuity between the corporeal and the heavenly. But his contemporaries, arguing for a discontinuity, thought that "physical causes can be dismissed altogether, and that it is fitting to explain astronomical phenomena only through astronomical methods" (cited in Kozhamthadam, 1994, p. 103). We must know that it was not until Newton that we had a mathematical proof and value of gravitation. For Kepler, the physics of the heavens was still a hunch. In the second edition, twenty years later, he proposes that "if we substitute for the word 'soul' the word 'force' then we get just the principle which underlies my physics of the skies in *Astronomia Nova*" (cited in Koestler, 1961, p. 51). It was an abductive inference that inclined (Peirce, 1976, 2.96) Kepler toward a hypothesis that, in turn, was made precise over the years. Koestler argues that to think of Heavens in terms of a physical hypothesis, to hypothesize a continuity is the most significant service of Kepler because:

> It represents primarily a search for one universal force law to explain the motions of planets — Mars in particular — as well as gravity and the tides. This breathtaking conception of unity is perhaps even more striking than Newton's, for the simple reason that Kepler had no predecessor.
>
> *1961, p. 72*

From Mean Sun to Real Sun

Contrary to the popular view, the Copernican model was not heliocentric in the true sense. His model had nothing to do with the sun particularly. Seeing the planet's motion as orbicular, the ancients thought of the planet's motions as circular (Kepler, 1609/1992, p. 115), and the centers of those circles were arbitrary points, not any bodies. Moreover, they resorted to souls to explain how a planet could maintain a perfectly circular orbit. Copernicus preserved this whole scheme. For him, all the planets were moving around an arbitrary point at the center of the universe. The sun just happened to be near that point in his model. His model was neither geometrically heliocentric, nor physically. Since MC, due to analogical comparison between heavens, spheres, and Trinity, that we have discussed, Kepler was convinced that it is the sun around which planets move, and not some arbitrary point. Where "Copernicus and Tycho followed Ptolemy, carrying over his assumptions," Kepler in his "*Mysterium cosmographicum*, take instead the apparent position, the true body of the sun, as my reference point." (p. 121) So he tested all the three hypotheses on the values adjusted for real sun and not the mean sun, hence the complete name of the concerned chapter in the *New Astronomy*, "On the equivalence of the hypotheses of Ptolemy, Copernicus, and Brahe, by which they demonstrated the second inequality of the planets, and how each changes when accommodated to the Sun's apparent motion instead of its mean motion" (p. 155, emphasis added). Kepler did not know of Mars' elliptical path at that time and lacked any evidence of sun's central position (Kozhamthadam, 1994, p. 147), but was convinced of his abductive hypothesis aided by analogy of sphere and his "metaphysical reasons." This hypothesis helped erase many doubts concerning measurements and prediction in Kepler's later work. The empirical data that came later in the form of an ellipse and its proposed explanation served as inductive confirmation of the hypothesis. It was this hypothesis of the real sun that, as its effects on future conduct, positioned our astronomer to think of ellipse and gravitation.

From Circle to Ellipse

Kepler's first law is about discovering Mars' orbit as elliptical. He, like others, was convinced initially that the orbit was a perfect circle, but it made it impossible to explain all of Mars's observed positions and made its orbit a pretzel. In the telescope's era, one can think of it as discovering a mathematical regularity by observing planets' motion at several points, but it was invented after Kepler had done his work. It seems intuitive that he might just have mapped the orbit on the measurements and discovered an ellipse, but one should know that he only had twelve measurements of Mars at that time (Wilson, 1972). Furthermore, discovering a circle is far easier than discovering an ellipse because the circle only has one perfect shape, while there can be infinite ellipses. This explains why Kepler failed seventy times before reaching the correct orbit and delayed his work considerably: "My first error was to suppose that the path of the planet is a perfect circle, a supposition that was all the more noxious a thief of time" (Kepler, 1609/1992, p. 417). He had to let go of the centuries-old commitment to geometrical perfection in the heavens. Of course, the Imperial Mathematician knew about the mathematics of ellipse, but he was "blind from desire" (p. 455) and held his commitment to the geometrical perfection of the circle.

The discovery of the ellipse explained Mars' positions without making its orbit a pretzel. This is an excellent example of a hypothesis eliminating doubt. It could not be an empirical curve-fitting procedure because the same measurements that Brahe used to fit Mars in a circular orbit, Kepler used to hypothesize an ellipse. He did not commit to this abductive flash without testing its validity, for then it would be fixing belief by the method of tenacity. However, it did affect his future conduct to test the hypothesis' different deductive formulations,

At a certain stage of Kepler's eternal exemplar of scientific reasoning, he found that the observed longitudes of Mars, which he had long tried in vain to get fitted with an orbit, were (within the possible limits of error of the observations) such as they would be if Mars moved in an ellipse. The facts were thus, in so far, a likeness of those of motion in an elliptic orbit. Kepler did not conclude from this that the orbit really was an ellipse; but it did incline him to that idea so much as to decide him to undertake to ascertain whether virtual predictions about the latitudes and parallaxes based on this hypothesis would be verified or not. This probational adoption of the hypothesis was an Abduction" (Peirce, 1974, 2.96, emphasis added).

Kepler's tenacity to the earlier hypothesis until a new one was conceived was not nonscientific or perverse in Peircean terms. From the point of view of pragmaticism, suspension of belief is not an option because our beliefs have practical effects. A scientist in the middle of an inquiry, motivated by irritation of doubt, cannot do this. Unless the inquirer has a new hypothesis, one should be testing different deductive formulations of the current hypothesis. Nor was his noncommitment to the new hypothesis irrational; as Peirce noted, he was just inclined to its possibility. Commitment to and the truth of any hypothesis go hand in hand as the probability of its truth increases with increasing inductive validity. If an inquirer agrees to or discards any hypothesis on the basis of a single test, then the inductive process has been reduced to the deduction.

Concept of Force

It was well known at the time of Kepler that the velocity of planets has an inverse relationship with their distance from the sun. Also, it was known that any planet moves faster in its orbit nearer the sun and slows down as it moves away. Still, Ptolemy and others, against observations, maintained that planets move at constant speeds (Kepler, 1609/1992, p. 145) and that the change in speed is an illusion – itself an abductive hypothesis. Now, the concepts of the Holy Ghost and the souls, in their own ways, had absoluteness and universality to them. It makes no sense that the souls become weaker away from the sun, or that the Holy Ghost tires. He argues:

> Either the souls which move the planets are the less active the farther the planet is removed from the sun, or there exists only one moving soul in the center of all the orbits, that is, the sun, which drives the planet more vigorously the closer the planet is…

> *cited in Koestler, 1961, p. 51*

Contrary to other astronomers, Kepler wanted a geometrical description of the heavens, and an explanation. Also, the broken idol of perfection of shapes and constancy of speeds made it more pressing to look for causes, so he was looking at "the matter dynamically, thought it must have something to do with causing the planets to move in their orbits" (Peirce, 1974, 6.604). In the *New Astronomy*, Kepler argues that "the sun is a magnetic body, and rotates in its space" (Kepler, 1609/1992, p. 385). The lengthy and convoluted discussion about the evolution of Kepler's concept of gravity is deliberately avoided here. It is appropriate to note that all his hypotheses in the process were aimed at sense-making, motivated by irritation of doubt expressed in the question *"A quo movementur planetae?"* However, it did not explain the variable speeds. He drew an analogy from a phenomenon that almost everyone experiences every day, that is, sunlight: "I reflected that this cause of motion diminishes in proportion to distance just as the light of the sun diminishes in proportion to the distance of the sun" (cited in Koestler, 1961, p. 51).

First, he hypothesized sunlight as a vehicle of that magnetic force but then dropped this hypothesis because it follows the force's inactivity at night as there is no sunlight then. He eventually broke away from the Aristotelean concept of force rooted in corporeal physics. This hypothesis eliminated the doubts caused by variable speeds of planets for Kepler and increased his confidence in the hypothesis of the sun at the center. This confidence increased further when he discovered a harmony in the form of his third law in 1619, about ten years after the *New Astronomy*. The ideals of beauty and perfection that he had to let go of in the *New Astronomy* when he resorted to ellipse were returned in the form of Kepler's version of the "golden ratio." For him, it was God's version of harmony that we humans try to find in our symphonies (Brackenridge, 1982).

Kepler's "Scientific Spirit"

For Peirce, nothing is more devastating for science than to use it to establish foregone conclusions (Peirce, 1974, 7.186). The perverse methods of fixing beliefs put one in an attitude to explain any experience to satisfy the existing beliefs. It seems that Kepler did precisely that. Peirce is well aware that every human being, because of different backgrounds and experiences, possesses a different set of preconceptions: "It makes no difference how imperfect a man's knowledge may be, how mixed with error and prejudice" (7.54). If it is not the so-called objectivity, what makes a good scientist and, more importantly, makes Kepler the "greatest reasoner who ever lived?" How could Kepler perform the "greatest piece of retroductive reasoning ever" if our prejudices make no difference?

Peirce wrote that it is a "moral quality," "without which a reasoner cannot escape fallacies, and that is a sturdy honesty of purpose" (Peirce, 1958, p. 256). One should ask oneself why one is doing science. One's logic of inquiry would follow from the answer to that question (Peirce, 1974, 7. 186). If the desire of an inquirer is to preserve beliefs, the inquirer should follow the method of tenacity or *a priori*.

If one wants to preserve the tradition, one should follow the method of authority. In different domains of life and at different stages of our intellectual development, we are entitled to opt for these different methods. Even in different stages of scientific inquiry, these methods are the most rational ways to proceed (Gilmore, 2003). These methods become rational if chosen consciously with certain ends in sight.

It was perfectly rational for Kepler to stick tenaciously to the circle unless a new viable hypothesis was available and, more importantly, until he had exhausted its all possible deductive formulations. As we have seen, Kepler knew that this hypothesis had the authority of centuries behind it. The argument that follows the geometrical perfection of the circle is an example of *a priori* reasoning. He used all the "nonscientific" methods to preserve his belief in the circle until a new hypothesis was found. What is this purpose with which an honest commitment prevents one from falling into perverse habits of thought? It is a "dominant passion of his whole soul must be to find out the truth in some department, regardless of what the color of that truth may be," it is not a method or a specific community that characterizes science but a spirit (Peirce, 1974, 7.54). If the inquirer pursues an inquiry with that spirit, it will provide a safeguard from becoming perverse in thinking.

The last question that remains to be addressed is about the nature of sense-making. What is sense-making in the context of science? How is scientific sense-making different from any other forms of sense-making? Associated questions would be why Kepler was not satisfied with the sense-making attempts of earlier astronomers? To put all these questions more precisely, what is science after?

Peirce praises Kepler's "fecund imagination" that "makes the clothing and the flesh drop off, and the apparition of the naked skeleton of truth to stand revealed before him" (Peirce, 1958, p. 255). Although his style is poetic here, his analogy of dropping off the clothes and flesh and the revelation of the skeleton is telling. Kepler was not after the appearances, the clothes, of individual planets. If he had approached the matter in this attitude, he might have turned toward Souls himself. Instead, he was after the skeleton of the whole system, taken together. The other three hypotheses were equivalent in explaining the position of Mars at individual points; the problem arises when we take the whole system together and try to make predictions. Mars had many clothes but no skeleton. Kepler was after its skeleton to put the right-sized clothes on it. What is this skeleton that our Astronomer revealed through his "fecund imagination?" While there are other types of imagination, scientific imagination "dreams of explanations and laws" (Peirce, 1974, 1. 48). The law, a regularity, would eliminate doubt by creating a habit, belief, of engaging with the world. For Peirce, like any habit, the habits of the universe, laws, are probable. It is the regularities in nature that science is after.

Ptolemy and other astronomers were subconsciously aware of this. Nevertheless, they tried to fit natural phenomena in the geometrical regularities they already had, instead of finding law, a regularity out in nature. The gravitational force, the Holy Spirit, the souls are higher-level habits that ensure the sustenance of lower-level habits of orbits as they are. Although the previous models adequately explained the instances in Mars' orbit, they failed in hypothesizing the regularities correctly. Metaphysically speaking, any instance is an existent, but regularity is real because "any habit, or lasting state that consists in the fact that the subject of it would, under certain conditions, behave in a certain way, is Real, provided this be true whether actual persons think so or not" (Peirce, 1992, p. 458). Although we are not in a position to pass any judgment on Ptolemy, Copernicus, Tycho Brahe, or any other scientist, any science or scientist that explains, or aims to explain, a given instance in itself without referring to any "Real" is not a realist, but a nominalist. Nominalists deny "the real existence of anything general, that is, of anything which wholly applies to many different things" (MS [R] 717). Doubt makes no sense in a nominalist universe because doubt and any need for sense-making presuppose that we expect something from the Universe. For Peirce, science, like any other sort of sense-making, is an antinominalist activity. Expectation entails regularity, reals that are not found in a nominalist universe. Kepler wanted to explain Mars' position and predict its course, which is only possible if that explanation is based on some regularity in the universe. An abductive hypothesis is an attempt at sense-making by making our habits of conduct congruent with the universe's habits.

QUESTIONS FOR DISCUSSION

1. One of the major criticisms by Kepler on the earlier theories was that they made the behavior of souls or the Holy Spirit erratic. It is not an empirical critique but rather metaphysical (souls) or theological (Holy Spirit). This begs the question about the relationship between the scientific and the extra-scientific. Like many others, Kepler is clearly a case where the extra-scientific helped propagate the scientific. Does that mean that the scientific must take into consideration the extra-scientific? The positive influence of extra-scientific on the scientific is not ubiquitous but not as rare as to discard it to serendipity. How can we conceptualize this relationship?

2. If the extra-scientific helped many scientists in their work, then one may deduce that it is empirical or, at least, has empirical elements. How do we understand the claim that science is based only on empirical evidence? Does this mean that scientific and extra-scientific are on a continuum between the nonscientific and the purely scientific? If this is so, is it only a matter of degree of being empirical? Or is there a qualitative difference between scientific and extra-scientific thinking?

3. Many of science's significant discoveries resulted from the tenacity of the scientist when the existing data or the scientific community were against her tenaciously held hypotheses. One can only know the fate of a hypotheses after the respective inquiry has been settled. Does that

mean holding onto a scientific hypothesis is akin to holding onto other tenaciously held beliefs? Another related question relates to the inquirer in this condition. What surety does a person have that any hypothesis will turn out to be true? Is it only hope? It might happen that someone's whole life's work would be deemed unscientific in the face of a changing paradigm. Personally speaking, is it worth striving for?

References

Brackenridge, J. B. (1982). Kepler, elliptical orbits, and celestial circularity: A study in the persistence of metaphysical commitment: Part II. *Annals of Science, 39*(3), 265–295.

Caspar, M. (2012). *Kepler*. North Chelmsford, MA: Courier Corporation.

Ferguson, T. S. (1989). Who solved the secretary problem? *Statistical Science, 4*(3), 282–289.

Gilmore, R. (2003). Peirce and perversity: The higher logic of the real. *Transactions of the Charles S. Peirce Society, 39*(3), 383–404.

Holton, G. J. (1988). *Thematic origins of scientific thought: Kepler to Einstein*. Cambridge, MA: Harvard University Press.

Kepler, J. (1981). *The secret of the universe*. Abaris Books. Original work published 1596.

Kepler, J. (1992). *New astronomy*. Cambridge. Original work published 1609.

Koestler, A. (1961). Kepler and the psychology of discovery. In E. P. Wigner (Ed.), *The logic of personal knowledge. Essays presented to Michael Polanyi on his seventieth birthday* (pp. 49–58). London: Kegan Paul.

Koestler, A. (2017). *The sleepwalkers: A history of man's changing vision of the universe*. London: Penguin UK.

Kozhamthadam, J. (1994). *The discovery of Kepler's laws: The interaction of science, philosophy, and religion*. Notre Dame, IN: University of Notre Dame.

Kuhn, T. S. (1957). *The Copernican revolution: Planetary astronomy in the development of western thought*. Cambridge, MA: Harvard University Press.

Osbeck, L. M. (2014). Scientific reasoning as sense-making: Implications for qualitative inquiry. *Qualitative Psychology, 1*(1), 34.

Peirce, C. S. (1877). The fixation of belief. *Popular Science Monthly, 12*, 1–15.

Peirce, C. S. (1958). *Charles S. Peirce, selected writings*. North Chelmsford, MA: Courier Corporation.

Peirce, C. S. (1974). *Collected papers of Charles Sanders Peirce*. Cambridge, MA: Harvard University Press.

Peirce, C. S. (1976). *The new elements of mathematics (Vol IV)*. Edited by C. Eisele. The Hague: Mouton Publishers.

Peirce, C. S. (1992). *The essential Peirce: Selected philosophical writings (Vol. 2)*. Bloomington, IN: Indiana University Press.

Robin, R. S. (1967). *Annotated catalogue of the papers of Charles S. Peirce*. Amherst, MA: University of Massachusetts Press.

Silva, A. P. R. C. D. F. (2007). *Metamorfoses do conceito de abdução em Peirce: o exemplo de Kepler* (Doctoral dissertation).

Teller, E., Teller, W., & Talley, W. (1991). *Conversations on the dark secrets of physics.* New York: Springer.

Whiteside, D. T. (1974). Keplerian planetary eggs, laid and unlaid, 1600–1605. *Journal for the History of Astronomy, 5*(1), 1–21. https://doi.org/10.1177/002182867400500102

Wilson, C. (1972). How did Kepler discover his first two laws? *Scientific American, 226*(3), 92–107.

2

GOETHE

A Person of Science

Michael V. Steder

Personal Preamble

Why Goethe? When considering a scientist to investigate for a case study, the long list of famous or impactful scientists throughout history provides many options. Johann Wolfgang von Goethe is rarely found on such lists and few people unfamiliar with the extent of his work recognize Goethe's scientific contributions and endeavors. Goethe's prominence is in the humanities, with renowned and era-defining literary work. If Goethe is not known as a scientist, why choose him to investigate?

I decided to investigate Goethe due to my personal interest and familiarity with his work. I first came upon Goethe in reading *The Sorrows of Young Werther* and *Faust* and was floored by the depth of feeling, rich personal expressiveness, and expansive range of diverse topics illustrated in these two works. My captivation led me to read as many of Goethe's literary works as I could find as I became familiar with his renown as one of the great European poets alongside Shakespeare and Dante. Because I only knew Goethe as a literary figure, I was astonished to discover his *Theory of Color* during my undergraduate studies. In *Theory of Color*, I found the same passionate, richly expressive, poetic descriptions as in his literary works, but instead directed to scientific inquiry. I went on to acquire an English translation of Goethe's *Scientific Studies* to become further acquainted with his works on natural science. I was impressed when I realized that he engaged in scientific inquiry throughout his adult life and made significant contributions in the areas of morphology, optics, and zoology. I find Goethe's scientific works captivating because he incorporates poetic language, artistic illustrations, and personal ruminations to describe and explain his

DOI: 10.4324/9781003276692-4

scientific findings. As such, I choose to investigate Goethe because he calls into question how artistic sensibilities and scientific rigor might complement and enrich one another.

Introduction

"In observing nature on a scale large or small, I have always asked: Who speaks here, the object or you? I also take this approach in regard to my predecessors and colleagues" (von Goethe and Miller, 1995, p. 308).

Goethe's statement from *Maxims and Reflections* voices a concern at the heart of both past and present scientific inquiry: What is the relation between the subject (i.e., the scientist) and the object of the scientist's inquiry? The traditional view regards legitimate scientific findings and conclusions as the result of objective, rationally determined, and methodical processes with effort to minimize the influence of subjective qualities and social or cultural processes (Singer, 1971). Accordingly, scientific inquiry is frequently touted as the exclusive means of objective discovery of general laws concerning the physical world. Historically, most studies of science are conducted in the domains of philosophy and sociology, providing valuable insights regarding conceptual and sociocultural factors involved in scientific activity and scientific knowledge. While less prevalent, psychological studies of science contribute to understanding the personal dimension of scientific inquiry but have primarily focused on cognitive processes and personality traits, narrowly limiting the scope of the psychological lens to individualist capacities and mechanical factors (Carré, 2019; Osbeck et al., 2011). Still, even these studies leave unacknowledged the interpersonal and intrapersonal aspects expressing the dynamics of the acting person engaging in scientific inquiry. Some recent studies have expanded exploration of the personal dimension and sought integrated accounts of science by focusing on science as active engagement, with an emphasis on scientists as persons, with all that this implies (Carré, 2019; Nersessian, 2012; Osbeck et al., 2011). In that tradition, this essay contributes to the psychological study of science by offering a multi-layered description of Goethe's scientific activity, drawing on autobiographical, biographical, and scientific writings.

I begin with a brief overview of the psychological studies of science to set the ground for the study of Goethe. Followed by an analysis of Goethe's activity of scientific inquiry, starting with a general overview of Goethe's personal and social history in conjunction with a review of his significant scientific works and contributions. Three of these works *On Granite*, *The Metamorphosis of Plants*, and *Theory of Color* serve as points of reference to reflect on Goethe's scientific activity. Discussion of each work considers his life history, disposition and interests, cultural and social factors, and forms of reasoning. With the social context of Goethe's scientific activity explored in greatest detail.

The Psychology of Science

The terms science and scientific are prevalent throughout this essay, however, as the history of science studies demonstrates, there is much controversy around defining these terms or their epistemic parameters. As Singer (1971) notes, to assist in demarcating factors relevant for psychological investigation science can be roughly viewed in two distinct categories: science as a specific form of cognition and science as a task. As a task, science proceeds via the active engagement of a person or persons seeking to explore and understand a particular observable phenomenon in a systematic fashion (Osbeck et al., 2011). This systematic exploration is not only cognitive but involves the person or persons' social roles, skills, affectivity, interests, commitments, and environmental contexts. As a task, scientific exploration is not an insular practice of a person or persons directing their particular traits to a sanctioned methodology but a coordinated interplay of activities across personal, cognitive, and social domains. To define science as a task is to regard the scientist as a person and science as a multidimensional activity embedded in environmental, personal, and social contexts (Carré, 2019; Nersessian, 2012; Osbeck et al., 2011). The description and analysis of particular activities is dependent on the research interest and object of focus of researchers conducting a psychological study. These studies of science, including case studies, do not offer an exhaustive description of activities but highlight different aspects of scientific activity relevant to the scientific task and context explored. In this essay, I explore a scientist's engagement with scientific inquiry through the course of his life and scientific writings in order to describe and analyze the multilayered interplay of personal, cognitive, and social domains of activity involved.

Biographic Overview

In order to describe and analyze Goethe's scientific activity with reference to specific personal, social, and cognitive aspects, a biographical overview is first needed. This overview is not extensive but gives a general impression of Goethe's life story in accordance with the major events, personal relationships, ideas, and beliefs which impact his scientific work. The information provided here is derived from *The Autobiography of Goethe: Truth and Fiction Relating to My Life* (von Goethe and Oxenford, 1969) and Boyle's (1991) *Goethe the Poet and the Age.*

Johann Wolfgang von Goethe was born in Frankfurt on August 14, 1749 to Johann Caspar Goethe and Catharina Elizabeth Textor. Goethe's father studied law and occupied a political position in Frankfurt affording his family a socially respectable and economically stable condition. His father placed a great deal of attention on Goethe to foster his son's interest in the liberal study of literature, history, language, geography, and art, especially music and drawing, employing numerous tutors. While his father sparked Goethe's intellectual and purposive ambition, his mother's love for the theater, open emotionality, and exuberant storytelling fostered a strong self-confidence in Goethe along with a disposition for theatric and poetic expression and imagination. In Goethe's own words: "My father gave me his build, his earnest conduct of life, my mother dear her happy nature and fondness for storytelling" (Boyle, 1991, p. 60). Goethe's early childhood parental influences mark an initial division in his interests, temperament, and pursuits which persists as an overarching theme over the course of his life.

Following his father's direction, Goethe began studying law in Leipzig in 1765 where he also continued to develop his interest in literature and painting. This illustrates the tension Goethe experienced in determining his occupational and educational pursuits. During his studies at Leipzig, Goethe attempted to reconcile this tension by setting his sights on attaining an academic teaching position "as the most desirable goal for a young man who intended to educate himself and contribute to the education of others" (Boyle, 1991, p. 62).

Goethe devoted himself with vigor to his college studies in both law and the humanities but was also avidly social among his peers. Effectively, this meant that he divided his attention in three directions. Spreading himself in these three areas led to overextension of Goethe's energies, triggering a physical and mental exhaustion that culminated a serious illness in 1768. This forced Goethe to return to Frankfurt for treatment and rest. During his recovery, Goethe became fascinated with Occult philosophy, particularly alchemy, and natural philosophy as a means to understand the material and nonmaterial qualities and essences of nature and human interiority.

Goethe reentered college at the University of Strasbourg, France, in 1770 to complete his law degree in 1773. During his studies at Strasbourg, the tension between Goethe's interests became more pronounced as he attempted to decide whether to pursue a life in the courts or one devoted to the arts in literature, drawing, and theater. During this period, Goethe befriended noted theologian, philosopher, and poet Johann Herder, who impacted Goethe's appreciation and understanding of classic literature and helped shape Goethe's own emotionally expressive poetry and literary works. In conjunction with Herder's theory of aesthetics, the emotional expressiveness of Goethe's writing along with his criticism and rejection of enlightenment ideals of rationalism and empiricism are instrumental contributions to the 1773 establishment and subsequent rise of the German literary and art movement known as "Storm and Stress" which promoted the importance of the freedom of individual, emotionally driven, subjectivity over the constrictions of rationalism.

Following the 1774 publication and success of *Sorrows of Young Werther*, Goethe's literary popularity helped to gain him a legal position in the court of Weimar in Germany. Goethe's decision to accept this court position was painful as it went against his father's desire for him to continue his legal education and pursue a loftier political title, marking an important instance in the determination of Goethe's identity and occupational direction apart from his father's trajectory. Goethe's first stint at Weimar as a court official of various positions lasted from

1775 to 1786. During this time, Goethe struggled to balance his personal, creative, and intellectual interests with his legal duties and was troubled by the tensions he experienced between adhering to the social norms of decorum and polite reserve with his desire for subjective expressiveness, fulfillment, and understanding. To avoid these tensions, Goethe turned his attention away from his literary endeavors and directed it toward a scientific investigation of geological formations occasioned by his appointment as a surveyor for the Ilmenau silver mines.

In 1786, Goethe decided to leave his time-consuming position at Weimar to journey across Italy in hopes of rekindling his personal passion for art and his emotional investment in literary and poetic works. Goethe traveled Italy from 1786 to 1788. Over the course of this trip, Goethe's understanding of the philosophic importance of aesthetics and artistic works changed from appreciating the intensity and complexity of emotional expression to holding a greater esteem for artistic balance and perfection. Goethe's shift during this period is from an emphasis on the importance of sensual and emotional fulfillment and expression toward a more detached, dispassionate investigation, description, and understanding of nature as the domain most readily demonstrating balance and perfection. Note that this is not a turn away from aesthetic interest but a shift in understanding how the aesthetic realm should be understood and engaged. During his time in Italy, Goethe focused his natural investigations on the structure of plants due to the availability and impact of the public gardens of Alcinous.

Goethe returned to Weimar in 1788, residing there until his death in 1832. While this period coincides with the French Revolution, Goethe was isolated from much of its direct influence. However, the revolutionary philosophical and social ideas circulating during this period are reflected in Goethe's immersion in scientific inquiry. This is particularly evident in Goethe's suspicion of methods of investigation founded on the enlightenment ideals of detached observation challenging traditional notions of the divine insight of ritual devotion, a suspicion which leads Goethe to critique Newton's theory of refraction. Goethe devoted most of his time corroborating and communicating on a variety of topics with scholars from the University of Jena. Discussions included natural science. The influence of Friedrich Schiller, a philosopher, playwright, and poet, is of particular importance for its duration and impact on Goethe's scientific activity. Goethe's conversations with Schiller led to the completion and publication of two major scientific works, *Theory of Color* and *The Metamorphosis of Plants*. Goethe's relationship with Schiller seems to have been a significant motivating factor for Goethe's scientific activity. Although Goethe continued his scientific inquiries until his death, his serious study of natural science ceased following Schiller's death in 1805.

Goethe's Major Scientific Works and Contributions

As the scientific publications printed in *Scientific Studies* attest, Goethe's scientific inquiries span across a variety of domains, most notably, geology, anatomy/

zoology, botany, and optics/physics. Yet, not only are his scientific contributions in these areas and others largely unrecognized due to his renown as a great poet, but he faced suspicions concerning the subjectivity evident in his investigative methods and his understanding of scientific epistemology (von Goethe and Miller, 1995; Boyle, 1991). However, as with his literary influence on the "Storm and Stress" movement, Goethe's scientific publications offered original discoveries, opened new lines of thought on various topics, and foreshadowed or influenced many subsequent scientists and scientific assertions (von Goethe and Miller, 1995; Boyle, 1991). Notable contributions from these publications include, but are not limited to, Goethe's discovery of the intermaxillary bone in humans (1820), the establishment of the notion of morphology (1790), an initial recognition of the physiological response to color (1810), and attention to a qualitative dimension in the study of natural phenomena (1832). Although Goethe published over thirty scientific works, I focus on three specific publications, *On Granite* written 1784, published 1877, *The Metamorphosis of Plants* written and published 1790, and *Theory of Color* written 1791–1807, published 1810, to explore the personal, cognitive, and social domains of Goethe's activity of scientific inquiry.

Goethe's interest in science may have originated in his fascination with nature in early childhood. As reported in his autobiography, at age seven Goethe reverentially celebrated the divinity of nature through a ritual practice involving the rising sun. Goethe saw the workings and objects of nature as the most explicit expressions of, and most direct access points to, God, and sought to worship God and nature together by arranging an altar of natural objects and burning aromatic pastilles with magnified sunlight. This childhood ritual illustrates the foundation of Goethe's approach to scientific inquiry and displays a precursor to his active participation in exploring, revealing, and describing the divinity of nature.

Goethe's 1784 *On Granite* is his first scientific work following the minor alchemical and chemical experiments he conducted in the 1770s. *On Granite* is a poetic rumination and depiction of the genesis, structure, and properties of granite as it relates to geological formations. In Goethe's own words, "[e]very journey into uncharted mountains reaffirmed the long-standing observation that granite is the loftiest and deepest-lying substance, that this mineral, which modern research has made easier to identify, forms the fundament of our earth" (von Goethe and Miller, 1995/1784, p. 131). Although this essay marks the beginning of Goethe's formal engagement with scientific inquiry, it was not intended as a stand-alone scientific work, but meant to be a section of a fictional story Goethe considered writing at the time. This point is worth mentioning because it shows that Goethe's scientific activity begins without a clear differentiation between scientific inquiry and artistic creation. This connection between the exploration and description of natural phenomena and artistic reflection and expression first presented in *On Granite* is a definitive feature of Goethe's science as a whole.

The Social Domain

Social processes that relate to the production of knowledge can be analyzed on a macro level, focusing on broad historical-cultural influences, including the intellectual zeitgeist of a place and a time. It can also focus on the micro level— the local and personal relationships that influence a scientist. Goethe's scientific projects can be analyzed with reference to both.

The zeitgeist of the late 1700s was characterized in part by widespread support for anthropomorphic and physico-theological approaches to understanding the natural world. A desire to offer an alternative to anthropomorphic explanations helped to organize and direct Goethe's inquiry detailed in *On Granite*. Goethe, largely due to his friendship with Herder and their discussions of metaphysical harmony, recognized a divine order in the formations, objects, and processes of nature. However, instead of anthropomorphizing, he focused on describing the harmonious aspect of nature as an effect of an emerging pattern of divinity expressing "that all things in nature have a clear relationship to one another" (von Goethe and Miller, 1995, p. 132). This emphasis on the divine and harmonious essence of nature continued to shape Goethe's scientific interests in botany, especially as portrayed in *The Metamorphosis of Plants*.

Questions about the relationship between the exterior forms of living objects and their interior processes or essence were central issues of consideration in discussions of nature in the late 1700s. Similar to the position he takes in *On Granite*, Goethe's project in *The Metamorphosis of Plants* (1790) is fueled by his disagreements with some of the then dominant ideas and theoretical premises concerning the laws that govern the structure and organization of the natural world. A prevalent idea at the time was the German concept of *gestalt* as structured form, the interrelation of organized parts into a whole. For Goethe, the problem with regarding natural organisms as relating to *gestalts* was the implication that the world is organized according to definitive developmental endpoints, sets of interrelated parts that enable and support the complex forms of living organisms. This consideration influenced the botanical studies of scientists such as Linnaeus in which the parts of a plant such as the leaves, bark, and buds of a tree are understood as supportive organs to the whole plant. Goethe's philosophic interest in metaphysical harmony through plurality, influenced by the ideas of Spinoza concerning the unity between nature, spirit, and God (von Goethe and Miller, 1995), is consonant with the relationship of parts to whole presented in the gestalt concept. However, Goethe disapproved of defining organic processes and natural organisms according to gestalts because of the static, fixed, universe implied by this view: "[w]ith this expression they [German thinkers] exclude what is changeable and assume that an interrelated whole is identified, defined, and fixed in character" (von Goethe and Miller, 1995, p. 63). The fluid and dynamic relations between parts are overshadowed by and subordinated to a predefined whole. In contrast, Goethe emphasized an ongoing flux and flow to natural organisms and their processes, as well as a plurality of forms over the life span. Goethe found this flux and flow and plurality most readily and simply displayed in the structure and growth of plants, the basis of his inquiry in *The Metamorphosis of Plants*. Through detailed botanical studies, he observed and described the multiplicity of natural forms and functions that "...produce a chain of creation without end" (von Goethe and Miller, 1995, p. 64). Influenced by Spinoza's emphasis on plurality as emblematic of God's infinitude, Goethe directed close attention to discerning the constitutive parts of plants to the minutest degree possible, as demonstrated by his observations concerning the parts that form a plant seed. The observations described in *The Metamorphosis of Plants* led Goethe not to reject outright but to build on the gestalt concept of parts to whole and propose the notion of morphology, illustrating how every plant structure is in a constant state of change. In Goethe's view, unities or wholes are formed not as endpoints of development but as moments of cohesion in an ever-unfolding process (von Goethe and Miller, 1995).

Similarly, Goethe's interest in the scientific inquiry of light and color was sparked initially by his critical evaluation of Newton's Enlightenment-influenced theory of light, which was widely and authoritatively upheld during the late 1700s and early 1800s. In his own studies of light, Goethe oscillated between a favorable and unfavorable attitude toward so-called objective methods of scientific inquiry. On one hand, Goethe took issue with Kant's refusal to grant human minds an adequate basis for understanding the essence of reality—reality in itself. In his criticism, Goethe favored Spinoza's acknowledgment of the power of human thought for grasping the harmonious unity behind the manifold complexity presented by objects, organisms, and processes (von Goethe and Miller, 1995, pp. 8–10). On the other hand, in keeping with Goethe's appreciation of perfection in classical art forms and their reproduction, he acknowledged the need to isolate objects of study for precise and controlled investigation. As such, Goethe's practice of scientific inquiry on display in *Theory of Color* (1810) is guided by his veneration for the mind's ability to attain truth from subjective experience as well as his appreciation for systematic and meticulous methods of experimental design. This dual commitment developed into a criticism of artificial conditions while acknowledging the need for careful and systematic study. Goethe criticized Newton's experimentation with refraction as artificial in its methods, displaying color and light phenomena of a special sort derivative of the experimental situation only. According to Goethe, the properties of color and light elicited by Newton don't reveal the essential qualities of color but instead reveal only one variety of color phenomena to the exclusion of others. The descriptive studies and experiments Goethe designed to investigate the nature of color demonstrate his intent to study color perception in the fullness and adequacy of the subject's experience with the immense complexity of color. In refutation of Newton's notion of complex light claiming color to be a property of a single substance, these studies culminated in *Theory of Color* and forwarded the idea that color is a plurality of kinds in three interrelated realms of reality: physiological, that is, properties related to the eye;

psychical, that is, properties related to reflective colorless objects; and chemical, that is, immanent properties of color that can be fixed or imparted to objects.

With regard to the micro level of social processes, Goethe's focus on granite rock, plant structures, and color phenomena is directly influenced by his occupational opportunities and personal relationships. Goethe's writing of *On Granite* coincides with his overseeing of the Ilmenau township's silver mines. Due to his occupational position, Goethe spent a great deal of time surveying mountains and mines in the Ilmenau region where his enthusiasm for mineralogy grew to a passion, eventually expressed in *On Granite*. Not only was Goethe influenced and directed by the dominant intellectual ideas of the time, he engaged these ideas by applying scientific inquiry to the particular environment he found himself in and the problems that presented themselves in those contexts. Goethe didn't choose to investigate geology because it was the best object or topic to prove his notion of the harmonious connection of nature. He immersed himself in the scientific inquiry of geology as a result of his direct experience with mining operations and regular expeditions to the Harz Mountains. And although Goethe found plants ideal objects of study to observe and describe morphology, his devotion to plants is also a result of local opportunities.

As noted, Goethe found the raw observational material for *The Metamorphosis of Plants* primarily in his visits to the public gardens of Alcinous during his stint in Italy following his departure from Weimar. Goethe regularly visited these illustrious gardens for his pleasure and occupationally as an artistic consultant for their cultivation. Goethe, ever keen to immerse himself in the fullness of the experiences presented to him, took his access to the Alcinous gardens as an opportunity, perhaps a fortuitous invitation from nature, to enter the discussion of natural forms and functions, observing and describing the various structures, transformations, and relationships in plant forms and development. Like the Harz Mountains, the Alcinous gardens inspired and shaped his investigations.

In regard to his *Theory of Color*, Goethe took advantage of the facilities at the University of Jena, and his personal interest in establishing and promoting the university's devotion to scientific studies, to experiment with the expressions and properties of color.

As his involvement with the Ilmenau mines, Alcinous gardens and University of Jena suggest, Goethe's immersion in geology, botany, and optics are not solipsistic enterprises motivated solely by personal passions. Goethe's mountain expeditions and mining surveys were not solitary endeavors but in an important way social activities, influences by his personal relationships with various scientifically and artistically minded colleagues associated with the University of Jena, such as J. C. Loder and G. M. Kraus.

Following his appointment at the court of Weimar, Goethe found himself socially isolated and artistically uninspired due to diminished interest in and even a cultural dismissal of the importance of sensual expression, with a corresponding insistence on emotional control and decorum. Scientific inquiry provided a way for him to find new connections and new emotional investment: "…Goethe's scientific turn brought him a German-wide common pursuit of the kind on which he thrived. It gave a new, technical, impetus to his drawing, in which the landscape inspiration had been languishing for years" (Boyle, 1991, p. 337). Not only were Goethe's artistic and intellectual passions rekindled in scientific inquiry, but he also broke out of his social isolation by forming relationships through collaborative activity and discussion of scientific ideas. During his trip to Italy, Goethe kept a correspondence with his intellectual companions and court personages from Weimar on topics including the expression of divine perfection in nature. These correspondences served to keep Goethe connected with others who shared his interest in understanding the natural order of reality while also providing Goethe a platform to develop his theoretic concepts of metamorphosis and morphology.

Goethe enjoyed the most direct involvement with the intellectual and scientific communities through his association with the scholars and activities of the University of Jena. Goethe mixed with a variety of professors and students which expanded and refined his interest in the humanities as well as the sciences. Goethe's most notable and most controversial scientific work, *Theory of Color*, was completed during his time at the University of Jena with the assistance and influential thought of Schiller. Schiller's concern for methodological precision and consistency helped direct Goethe's experimental designs and argumentative style, culminating in the *Theory of Color*. Without Schiller's involvement, Goethe's exploration of color would likely remain a contemplative exercise or a series of disconnected observations lacking theoretic assertions (Boyle, 1991; Steiner and Clemm, 2008).

The Personal Domain

The personal domain of science or any other activity consists of the intellectual style, perceptive and emotive dispositional qualities, and manners of temperament, experience, values, commitments, and creative expression. Goethe's life, as depicted in both his autobiography and Boyle's biography, reveals a search for balance and a desire to illuminate the unity of nature and humanity. These are reflections of Goethe, the person through the entirety of his life. As such, they inhere his science as much as any other activity.

Goethe's search for personal balance can be seen in early childhood as he attempted to devote equal attention to his father's educational demands and his own creative and artistic sensibilities and skills. Goethe regarded himself as split between two paths which continued into his early college years and where he attempted to balance his academic studies with his social and artistic endeavors at the cost of his health. Upon returning home in 1768 to recover, Goethe was introduced to the scientific investigation of nature through the works of alchemy presented to him by Metz and von Klettenberg. In alchemy, Goethe found a balance that eluded him in his personal life. Alchemy presented a unified system for exploring and understanding the reflection of human life and spirit in the balance of material and nonmaterial nature. This alchemical emphasis on connectivity and balance is a guiding undercurrent in *On Granite*. Moreover, the content in *On Granite* is a poetic description of the forms and qualities of geological structures, but with subjective rumination or insightful contemplation that centers on mystical and spiritual revelations connecting the human spirit with the nonmaterial aspects of nature. As Goethe states:

> [i]n this moment, when the inner powers of the Earth seem to affect me directly with all their forces of attraction and movement, and the influences of heaven hover closer about me, I am uplifted in spirit to a more exalted view of nature. The human spirit brings life to everything, and here, too, there springs to life within me an image irresistible in its sublimity.
>
> *von Goethe and Miller, 1995, p. 132*

The poetic tone enables Goethe to describe the personal impact of his observations of rock formations in the Harz mountains. However, this poetic description also conveys a contemplative insight concerning the balance of nature's forces. This insight is a spiritual revelation expressing the human spirit's ability to encounter and disclose the divine order and movement of nature, a notion also central in alchemy. As the above quotation demonstrates, Goethe's curiosity concerning granite extends beyond descriptions of rock formations and emotive impressions to considerations of balance in the universe, culminating in an insight into the unified essence of nature connecting the human spirit with the divine.

As noted, the scientific activity resulting in *The Metamorphosis of Plants* coincides with Goethe's trip to Italy prompted by fatigue from his duties as a member of the Weimar court and a lull in artistic inspiration. On visiting the Alconian gardens of Italy, Goethe not only recuperated through the tranquility of his surroundings, but he also developed an intent to balance his artistic sensibility with insights into the divine perfection that connects the inner essence of plants to their outer structures.

During his time in Italy, Goethe's artistic interest shifts from the richness of emotive expression to the appreciation of skillful expressions of technical precision and balanced harmony of content and form. This artistic interest found an outlet in Goethe's scientific activity as an intention to recognize and reproduce, in both image and word, the minute details of plant structures to convey the harmonious essence of their multiphasic genesis. As with *On Granite* the intention in *The Metamorphosis of Plants* is directed beyond descriptive renderings of forms to the unifying process guiding the formations of various plant structures. For Goethe, distinguishing and reproducing plant structures as isolated inventories is a preliminary step for categorization, but categorization is not the goal of scientific inquiry. Goethe's inclination toward recognizing a divine order to nature leads to his intention for a higher-order insight into the organizational processes that harmonize the unfolding of specific plant structures such as leaves, stems, and buds into the plants' unified form. This higher-order intention in the employment of artistic precision is emblematic of the personal considerations of balance and connectivity shaping Goethe's scientific activity.

Prompted by his conversations with Schiller and the popularity of enlightenment methods of investigation growing in the natural sciences, Goethe's concern in *Theory of Color* turns to the role and importance of the subject (the researcher) conducting an experiment. *Theory of Color* is directed by Goethe's intent to demonstrate the necessary connections between the subject experiencing or observing a phenomenon and the object observed or experienced. For Goethe, scientific activity, like artistic creation, is a means to encounter and demonstrate the connection between subject and object, unifying the human spirit with the divine. In *Theory of Color*, Goethe's intention to connect subject and object is revealed by his emphasis on the three interactive qualities that define color. Goethe's experiments with color and light are designed with the intent to reveal a harmonious interplay of objective qualities (e.g., the qualities of light) with subjective qualities (the properties of the visual system). Goethe's experiments and the conclusions he draws from them demonstrate color as a relational phenomenon balanced between objective and subjective qualities.

The Cognitive Domain

The cognitive domain of science refers to types of reasoning processes utilized in conducting scientific inquiry. In *On Granite*, Goethe's descriptive reasoning process begins with firsthand observational experience from which his descriptions proceed with precise detail; they are then followed by reflection on the commonalities between particular observations. That is, the careful descriptions are followed by contemplation through which he gradually makes inferences toward a general conclusion and contemplation till a general conclusion is reached via inference.

This inductive reasoning is also utilized in the descriptive precision of plant structures in *The Metamorphosis of Plants* to identify the multitude of parts that define specific plant structures and their development. In distinction from the anthropomorphic theoretic tendencies of his time, which began with a theory concerning the purpose of plant development relating to human needs, Goethe began with patiently and meticulously observing, describing, and meditating on the genesis of plants themselves. This allowed him to devise a ground-up theory of plant metamorphosis centered on a plant's unfolding essence in conjunction with its external, that is, environmental context.

Goethe's inductively driven descriptive reasoning is further refined to a systematic method in *Theory of Color*, his most thorough scientific project. Unlike Newton, Goethe's optical theory results from a variety of different experiential conditions with light and color after much careful contemplation of numerous descriptions. In *Theory of Color* Goethe is not concerned with testing a hypothesis to determine and measure the causal properties of light. Goethe's reasoning is directed to experiencing and describing as many different instances and conditions of color as possible in order to arrive by contemplation of such multiple forms, at general conclusions concerning their essential features.

Despite following the cannons of inductive reasoning, Goethe's descriptive process is not a mechanical, detached procedure. It reflects the insights of a uniquely gifted and sensitive person, with details expressed in emotive, sensual, and contemplative depth, and consistent with spiritual and philosophical awareness. The artistic factor of Goethe's reasoning takes it beyond a solipsistic inductive process, exposing it as a relational, immersive, engagement. Goethe's poetic ruminations are not detached but implicate him in the scene of the observation as an active participant. Even in *On Granite*, he does not simply describe geological objects but is moved and inspired by them, feeling a connection to the geological structures he encounters. Similarly, Goethe's descriptions of plant structures not only provide patterns for his theory of metamorphosis, but his contemplation on the numerous variety and harmonious interplay of plant genesis enhanced his awareness of an ever-morphing essence in the divine order of nature. In his *Theory of Color* Goethe does more than present a theory of optics in opposition to Newton. He finds in the experience of, and experimentation with, color phenomena a harmonious interaction between subject and object—a relationship between the qualitative impressions of a human with the essential qualities in encountered objects.

Conclusion

In this essay I offered a psychological study of science by analyzing Goethe's scientific inquiry across social, personal, and cognitive domains. This was accomplished by focusing on specific aspects and processes of these domains as they relate to the content of three of Goethe's scientific writings. The analysis of Goethe began with an overview of his life and scientific works to provide a background against which to discuss the various aspects of his scientific activity. Goethe's *On Granite*, *The Metamorphosis of Plants*, and *Theory of Color* served to highlight his scientific activity for the purpose of elaborating specific personal, social, and cognitive aspects and processes. The exploration of these works foregrounds Goethe's scientific inquiry as an active engagement with, but also critique and refinement of dominant ideas about nature. I also explored ways his scientific studies were affected by the fortuitousness of Goethe's occupational positions and social interactions. Finally, my analysis also describes Goethe's scientific activity as a kind of mystical and spiritual practice based on his search for balance and a search for unity between the divine, nature, and humanity. Therefore, the cognitive processes of scientific inquiry Goethe employed, though inductive, also incorporate a relational and emotional encounter with the objects he studied and, artistically rendered descriptions employed for the purpose of making more general scientific conclusions.

QUESTIONS FOR DISCUSSION

Three questions helped direct this inquiry:

1. How does Goethe's search for spiritual, personal, and occupational harmony shape his scientific activity?
2. How might artistic sensibilities and expression contribute to scientific knowledge?
3. How can science be understood as a rigorous procedure while acknowledging the important role of subjective experience in producing knowledge?

References

Boyle, N. (1991). *Goethe: The poet and the age.* Oxford: Clarendon Press.

Carré, D. (2019). Towards a cultural psychology of science. *Culture and Psychology, 25*(1), pp. 3–32.

Nersessian, N. J. (2012). Engineering concepts: The interplay between concept formation and modeling practices in bioengineering sciences. *Mind, Culture, and Activity, 19*(3), 222–239. https://doi.org/10.1080/10749039.2012.688232

Osbeck, L. M. Nersessian, N. J., Malone, K. R., & Newstetter, W. C. (2011). *Science as psychology: Sense-making and identity in science practice.* New York: Cambridge University Press.

Singer, B. F. (1971). Toward a psychology of science. *American Psychologist, 26*(11), 1010–1015. https://doi.org/10.1037/h0032255

Steiner, R. & Clemm, P. (2008). *Goethe's theory of knowledge: An outline of the epistemology of his worldview.* New York: Anthroposophic Press.

von Goethe, J. W., & Miller, D. (1995). *Goethe's collected works: Scientific studies.* Princeton, NJ: Princeton University Press.

von Goethe, J. W., & Oxenford, J. (1969). *The autobiography of Goethe: Truth and fiction relating to my life.* New York: Horizon Press.

3

THE PROCESS OF MENDELEEV'S DISCOVERY

A Multidimensional, Relational Perspective

Ahmed Asad

Personal Preamble

This case study was written for a class titled Psychology of Science offered by Dr. Lisa Osbeck at the University of West Georgia. The purpose of this class, and this paper, was to explore the lives of actual scientists from a perspective that is often underappreciated, if not blatantly ignored. This perspective chiefly consists not in looking at the scientist's discovery from a strictly logical or scientific angle (where the logical form and scientific content of the discovery are held central), but from the broader sociological, psychological, and intellectual context.

Why did I choose Dimitri I. Mendeleev? For the past few years, I have been interested in American pragmatism, especially the works of Charles Sanders Peirce. Peirce claims that in the history of science, the best example of inductive reasoning can be found in Mendeleev's discovery of the periodic law. For me, if Peirce, a trained logician, held Mendeleev in such high regard, it seemed worthwhile to explore the process of Mendeleev's discovery not only in greater depth, but also greater breadth—investigating this process from an intellectual, psychological, and sociological perspective.

Every inquirer invariably brings, implicitly or explicitly, certain assumptions to the process of inquiry. In writing this chapter, I am committed to "Peircean Realism." Realism, as explicated by Peirce, among other things, means that relations are real and not merely projected onto the phenomenon by the inquirer. Indeed, Mendeleev himself can be seen as committed to a certain kind of Realism as he was convinced that the disparate elements (as viewed during his time) were somehow unified—indeed this conviction lay behind his discovery of the periodic law. My commitment to Realism lies in the conviction that the *content* of

DOI: 10.4324/9781003276692-5

Mendeleev's discovery is distinct from, yet inseparable from, the historical *form* taken by his discovery.

Purpose

Science is particularly interested in moving from the known to the unknown. If there is continuity between the known and the unknown, or, past, present, and future, a critical examination of the past affords a more informed picture of the present and promises more realistic expectations and ideals for the future. What does the exploration of Mendeleev's case study offer in this regard? Mendeleev's case study reveals the conditions under which the discovery process is made possible and conducive. An exploration of these conditions has great pedagogical value, especially pertaining to the initiation and training of young inquirers. A deeper look at these conditions and the potential pedagogical value they hold will be discussed toward the end of this paper. The purpose of this exploration is to inquire into the conditions in which an actual historical discovery was made, with the hope of identifying the pivotal moments which made this discovery possible. It is hypothesized that these pivotal moments hold invaluable pedagogical lessons for the training of future scientists, including psychological scientists. More importantly, if a critical inquiry into the actual conditions and local context of discoveries is missing, it robs us of the ability to extract pedagogical lessons vital for the training of emerging scientists, and hence, a more efficient movement of science.

Introduction and Motivations

Dmitri Ivanovich Mendeleev (1834–1897) was a Russian chemist best known for his discovery of the periodic law in the nineteenth century. His discovery chiefly consisted in identifying and delineating the periodic relations between the chemical properties and atomic weights of the elements—the periodic table is the systematic, iconic representation of these relations.

Mendeleev developed his career in chemistry at a time when major social and economic changes were taking place in Russia, a time just after the Crimean War referred to by many historians as "the Great Reforms Era" (Kaji, 2018, p. 224). This era of reforms in Russia was paralleled by a period of significant and foundational changes in chemistry as well. Mendeleev was deeply interested in both the welfare and development of his Russian homeland as well as the conceptual developments and problems plaguing chemistry at the time. In fact, Mendeleev's interest and work in chemistry can be regarded in light of his patriotism, as Kaji (2018) mentions that "the emergence of a new generation of chemists in Russia, who were eager to engage in original laboratory work in chemistry, was an important background to Mendeleev's activities during this period" (p. 224). Mendeleev, along with other Russian chemists at the time, was interested not only in the conceptual conundrums of chemistry, but viewed these conundrums in light of the practical problems faced by Russia. For Mendeleev, the question of how chemistry could contribute toward ameliorating the Russian condition was intimately related to the question of ameliorating the problems faced by chemistry itself. For him, the way to simultaneously deal with these distinct yet related practical and theoretical problems was via pursuing a career not only as a researcher in chemistry but as a teacher in a pedagogical institute. By the help of his teaching position, he (and other chemists) could deal with both the theoretical problems of chemistry while at the same time producing educated professionals able to deal with the industrial needs of Russia. Mendeleev's deep interest in both these areas is further manifested by his proposal to relevant Russian scholars to translate important non-Russian books on technology and agricultural issues to Russian as well as being a founding member of the Russian Chemical Society (Kaji, 2018). Furthermore, Mendeleev decided to write his *Principles*, the book he was writing when he discovered the periodic law, precisely because he could not find any good introductory book on chemistry in Russian which he could to recommend to his students. As Kaji (2018) mentions, "Mendeleev's *Principles* was the culmination of his work to help satisfy his country's demands during that period, both theoretical and practical." (p. 225). For Mendeleev, these issues were so intimately related that the "practical problems of economic development" in Russia, which they even made their way into the textbook he was writing (Bensaude-Vincent, 1986). To summarize, Mendeleev was not merely embedded in a context where both the foundations of Russia and chemistry were being laid; his activities reveal that he thoroughly embodied both these movements, as for him, both were closely intertwined. In other words, Mendeleev was truly in synchrony with the *spirit of the times*, or to quote Bensaude-Vincent:

> Mendeleev so wholeheartedly shared the faith of his epoch in the progress of science and technology that he could not separate the future of chemistry from the future of Russia: "What has been sown for the field of science will grow up for the people' welfare" he said, encouraging young Russian chemists.
>
> *Bensaude-Vincent, 1986, p. 3*

Intellectual Zeitgeist

Mendeleev's faith in the progress of science and general patriotic motivation is clear; his discovery can be seen as the culmination of both. In light of his original contribution toward chemistry, a deeper look at the state of affairs in the discipline of chemistry and its conceptual standing at Mendeleev's time is warranted. Chemists had been interested in discovering new substances long before Mendeleev; however, with the advent of new technologies such as spectral analysis in the 1860s, the efficiency of these discoveries significantly increased, resulting in a plethora (around 60) of newly discovered chemical substances

(Bensaude-Vincent, 1986). Even though the theoretical foundations of chemistry had undergone significant changes as well, especially with the emergence of the classical organic structural theory, *the* question plaguing chemistry at the time pertained to the relationship between the discovered substances (Bensaude-Vincent, 1987). To reiterate, although chemistry advanced in two of its most important professional undertakings—discovery of new substances and conceptual development—*the* theoretical issue of the relationship between the chemical substances was still far from settled. It was precisely when this conceptual problem was most stark that Mendeleev wrote his *Principles*. The important point is that Mendeleev, while writing his *Principles*, faced the exact same theoretical problem plaguing chemistry, but from a practical and pedagogical point of view. As he was writing an introductory textbook on chemistry for his students, he had to present an overview of all the discovered elements, yet, as there was no general principle relating all the elements at the time, he was troubled with the issue of how to organize the elements in his book. There were few known families of elements (groups) at the time which he used to arrange some elements in the first part of his *Principles*, but the problem manifested itself when he had exhausted the known groups and was bewildered about how to organize the remaining elements (Kedrov, 1966).

At this point, it is instructive to consider where mainstream chemistry stood at this time, particularly in relation to the above-mentioned problem, and the conceptual development of Mendeleev within this context. The French chemist Antoine Lavoisier had provided chemistry with a new vocabulary, a rational nomenclature describing the chemical composition of each substance (Bensaude-Vincent, 1986). According to Vincent "the new language did not only change the vocabulary used by chemists but also their theories and even their practice" (p. 4). Central to this theoretical change was the notion of "simple substance"—something which could not be further decomposed by any known chemical means. What is of primary importance is that this simple substance was "defined in a purely negative manner" (much like the indivisible particle of physics) (p. 4). For Bensaude-Vincent (1986), this notion is central to understand the development of chemistry in the nineteenth century. Precisely, as nothing positive was predicated of this simple substance, the negative void encouraged attempts to fill it via speculation. The nature of this speculation was reductionistic simplification—speculation about an ontological unity behind the empirical substance (see Slife and Williams, 1995, for more on "simplicity," "unity," and "reductionism" as they are used here). Two traditions of speculative reductionism were borne out of this negative simple substance: William Prout's unitary (monistic) reductionism and John Dalton's atomic (pluralistic) reductionism.

William Prout's unitary reductionism, more commonly known as "Prout's unitary hypothesis," was presented in 1810 (Bensaude-Vincent, 1986). According to this hypothesis, all simple substances were, in actuality, reducible to a single "primary matter." Originally, this primary matter was hypothesized to be hydrogen, and given the name "protyle," but ultimately, it was considered to be an unknown element constitutive of all simple substances. Ironically, Dalton's atomic hypothesis (pluralism) actually served to lend support to Prout's unitary hypothesis (monism) by virtue of its speculative and reductionistic character. Thus, the nineteenth century, from the point of view of chemistry, was dominated by the "unitarian mainstream" aspiring to "ground their unitarist faith on a network of relationships between elements" (Bensaude-Vincent, 1986, p. 6). More clearly, this aspiration was directed toward answering *the* problem plaguing chemistry at the time; the problem of how to unify chemical substances, as mentioned above.

Yet, what is important to point out here is not the yearning for an answer per se, but the *kind* of answer the unitarian mainstream was looking for. It may be argued that only a particular kind of unity (answer) was possible within the unitarian framework. What was the nature of this unity? As the unitarians hypothesized a primary matter, the unity they were looking for was of the nature of a classification, where all classes, having similar members, are related solely on the basis of common origin. More precisely, they were in search of a genealogical classification of substances in which similar chemically similar substances could be arranged into isolated families known as groups. This was a unity on the basis of origin, where groups were related merely on the basis of their origin—primary matter. According to Vincent, almost all of Mendeleev's precursors were proponents of this unitarian mainstream, which serves to show how thoroughly Mendeleev was embedded in unitarian conceptions. Yet, Mendeleev's trajectory was in the opposite direction, as Vincent states:

> He had the conviction that chemical elements were actually individual, and that they would never be divided or converted into another element. Until his death, he fought against all the attempts to reduce their number, and was bitterly disappointed when he realized that his own discovery had turned to be a chief argument in favour of Prout's hypothesis.
>
> *1986, p. 7*

Once again, it is ironic that Mendeleev's elemental individualism was used to support the reductionism it revolted against, just as Dalton's atomic hypothesis was used to support Prout's unitary hypothesis. At first glance, it appears that Mendeleev's elemental individualism and Dalton's atomic hypothesis go hand in hand, indeed this is the position taken by Bensaude-Vincent (1986). However, Kaji (2018) argues that this is a misreading of Mendeleev's position, in light of Mendeleev's conceptual trajectory. In order to understand this apparently subtle, yet vital distinction, it is important to first understand the difference between "simple substance" and "element" (Bensaude-Vincent, 1986, p. 14).

Mendeleev's Conceptual Trajectory

Why was Mendeleev opposed to atomic theory? According to Kaji (2018), Dalton's atomic hypothesis rested on "the law of definite proportions," but Mendeleev's early research on indefinite compounds in the 1850s reveals that he was starkly aware of the departures from or exceptions to the law of definite proportions. As Kaji (2018) states:

> Mendeleev was always cautious about atomic theory, and he made this clear in a lecture on theoretical chemistry published in 1864: In fact, while the atomic theory was strongly supported by the law of definite chemical compounds, it was also challenged by the so-called indefinite compounds.
>
> *p. 224*

Herein lies the subtle distinction alluded to above, as Kaji (2018) states, "It is reasonable to suppose that he [Mendeleev] refined the concept of the elements to incorporate individual chemical entity without employing the notion of atoms because of the supposed limitations of atomic theory" (p. 227). Before Mendeleev, the meaning of element was considered to be the above-mentioned simple substance. Even though Dalton had proposed his atomic hypothesis, tying elements to atomic weight, and hence giving a positive definition to elements (as simple substances), "this first connection did not generate a new concept" resulting in the "unavoidable pluralism of simple substances" (Bensaude-Vincent, 1986, p. 5). However, according to Kaji (2018) Mendeleev's key insight was that atomic weight did not belong to simple substances or compounds, but to elements (in contradistinction to both substances and compounds). Elements still remained something abstract, but, for Mendeleev, the only intelligent way to talk about these abstract elements was not via reductive speculation—primary matter or atomic form—but via their empirical effects—atomic weight and chemical properties. This can be considered Mendeleev's primary and initial abductive hypothesis (further explicated later in this article), which made the discovery of the periodic law possible.

In light of Mendeleev's discovery, what is the retrospective significance of the discussion above? Mendeleev's opponents were interested in unifying the elements; the nature of this unity is mentioned above. Yet, Mendeleev was also interested in a general unifying principle. What then distinguished Mendeleev's unity from his opponents? The nature of both unities is betrayed by their respective notion of elements. For the opponents, the elements were simple undecomposable substances, behind which lay an ontological unity—primary matter or atoms. In other words, they were searching for a fundamental unity behind the substances, toward their origin. On the other hand, Mendeleev's unity consisted of the empirical effects of abstract elements. Stated differently, Mendeleev was neither interested in homogeneous matter nor in definite forms (both of which he considered to be speculative) behind empirical substances. Rather, he was interested in the concrete and sensible effects of abstract but real elements. Mendeleev was therefore not interested in the genealogical classification of substances based on their similarities but was interested in the periodic regularities observed in the effects of elements, ultimately revealed in his periodic law. Furthermore, the unitarian orthodoxy

> could hardly do more than identify a few isolated groups or families. Mendeleev, on the contrary, wanting a strict general law, focused first on *dissimilar* [emphasis added] elements such as sodium and chlorine; he built thus a global framework before going further on to examine more detailed relationships. The idea of comparing the two extreme groups of halogens and alkaline metals happened to be the key of his success for it revealed regular differences in the atomic weights values between two neighbouring elements.
>
> *Bensaude-Vincent, 1986, p. 10*

As Mendeleev was interested in a truly general law, he was equally interested in the relationship between dissimilar elements, whereas the unitarians could merely group together some similar elements utterly unrelated to other groups of dissimilar elements. In other words, Mendeleev was interested in the relationship between *all* effects in the form of a law and not merely classification which would preclude relationship between groups (dissimilar elements) except in relation to their speculative origin. Simply put, Mendeleev was interested in real relationships, not speculative reductionism. Reducing elements to a single substance was to compromise not only on their individuality but also the generality Mendeleev yearned for, as "no general relation is possible between things unless they have some individual character in them" (Mendeleev, 1904, as cited in Campbell, 2017, p. 95). The reason that the orthodox unitarians could not predicate genuine individual character to the elements, and hence could not move toward a general law, was that "they were not able to evolve the conception of any other possible unity in order to connect the multifarious relations of matter" (Mendeleev, 1889, as cited in Campbell, 2017, p. 95)

Charles Peirce's paper, "How to make our ideas clear" (1878), helps us position Mendeleev's trajectory in relation to mainstream chemistry at his time. The gaze of the mainstream was fixed toward the origin, definitions, and classification of the substances which can be seen as relying solely on what Peirce calls the first and second degrees of clarity, that is, familiarity with the substances and their precise abstract definitions. Mendeleev's attempt to clarify the meaning of the elements can be seen as oriented toward the third degree of clarity, which characterizes Peirce's pragmaticism as stated in his infamous pragmatic maxim, "Consider what effects, that might conceivably have practical bearings, we conceive the object of our conception to have. Then, our conception of these effects is the whole of our conception of the object" (Peirce, 1878). It is precisely Mendeleev's focus on sensible effects that reveals his pragmaticist orientation and sets him apart from the speculative orientation of the mainstream. However, Peircean pragmaticism is not shallow utilitarianism, where any conception is used insofar that it works; rather, the issue of truth is central to his pragmaticism. Accordingly, it is interesting that Mendeleev actually used valence theory (tied to atomic theory which he opposed) to organize the first part of his principles, but as Mendeleev mentions in his own writings, this was merely to "simplify a system" (Kaji, 2018). In other words, his use of valence theory was out of its *utility*, whereas his yearning and search for a more secure principle is another sign of his pragmaticist orientation toward truth. Furthermore, for Peirce, the evolutionary trajectory toward truth, via the process of inquiry, is characterized by a simultaneous increase in diversity and regularity (CP 1.174). Therefore, attempting to reduce away genuine diversity (individual elements) will inevitably thwart the concomitant movement toward regularity (periodic law). Mendeleev's conception of elements as "individuals where diversity exists" (Campbell, 2017, p. 95) and his yearning toward a generality without compromise on individuality again shows his pragmaticist trajectory.

Another sign of Mendeleev's pragmaticism consists of the novel and concrete questions his discovery generated (see Kaji, 2018 and Campbell, 2017), as for Peirce, genuine inquiry opens further concrete lines of inquiry. However, it has to be conceded that Mendeleev could not fully escape the influence of his predecessors and ultimately tried to pour "the new wine [concrete effects] in the old bottles [perfect unity]" which, according to Dewey (1920, p. 51), characterizes modern thought in general. The old bottles consisted of his idea of what a natural law is—perfect without exceptions—whereas for Peirce, chance is an irreducible element in the universe and is at play in every natural law. Perhaps, this is the "internal" reason that Mendeleev's own discovery, even though borne out of his antispeculative and pragmaticist trajectory, helped strengthen the same atomism he opposed, and was ultimately seen as espousing the "positivist ideal" of empiricism (Bensaude-Vincent, 1986). However, before proceeding, a word of caution is in order. Vincent presents Mendeleev as an atomist, whereas Kaji presents him as an anti-atomist. Campbell (2017) states that "Mendeleev's stance on atomism is more difficult to fix..." (p. 59) as revealed both by his apparently contradictory statements on atoms and the fact that scholars are divided on this issue. In light of the discussion above, it may be concluded that, like the issue of unity, Mendeleev was not opposed to unity *per se* but a particular kind of unity. His position on atomic theory can also be interpreted in this manner, as unease with a particular kind of atomism (Daltonion), but without the ability to articulate an alternative theory. Indeed, Peirce was also opposed to Daltonion atomism but not atomic theory in general, as evidenced by his movement toward a conception of an alternative atomic theory which was dynamic in character (Campbell, 2017). To summarize, Mendeleev's hostility toward the orthodox conceptions of primary matter, atoms, elements, and generality all centered around a revolt against a particular conception of unity which these conceptions entailed. It is argued here that Mendeleev's particular notion of unity is intimately related to his later discovery of the periodic law; in fact, it may be that he was able to see such a unity precisely because he was looking for it. Yet, Mendeleev was so thoroughly embedded in the metaphysics of his time that he could not fully escape its influence, as betrayed in his conception of laws as being absolutely fixed.

The Day of the Discovery

We are now in a position to better appreciate the day and context of Mendeleev's discovery. Seldom is a discovery so thoroughly, though inadvertently, reported as Mendeleev's. Kedrov's (1966) archival analysis of the day of Mendeleev's discovery is an invaluable resource in this regard. According to Kedrov, Mendeleev had just finished the first chapters of the second volume of his *Principles* in early 1869. As mentioned earlier, he had used valence theory as a means to simplify his organization, yet due to his unease with atomic theory, he was both in search of a more secure principle and bewildered about how to organize the rest of the elements in his book. Finishing the first chapters of his second volume meant that the issue of an organizing principle, especially for the rest of his book, was most salient at this time for him. Mendeleev decided to take a break from writing and teaching in order to inspect the organization of some dairy artels in Russia for a voluntary economic society with which he was involved. In light of these facts, we can see two, seemingly contradictory, psychological trajectories at play: a yearning to find the answer to a salient practical and theoretical problem and the requirement to bracket his work in order to fulfill a social responsibility. The tension between both trajectories would have been most stark the day he had to leave for his inspection, 17th February, which was also the day of his infamous discovery of the periodic law. Furthermore, "from the standpoint of the activities on which Mendeleev planned to spend this day as well as the next eleven days" it is clear that Mendeleev's discovery "came about completely unexpectedly" as it shows that he "did not at all intend…to take up any large scientific question that might consume much time for their solution or that could detain him in Petersburg" (Kedrov, 1966, p. 20). Before diving deeper into the psychological factors pertaining to the discovery, a closer look at the content and logical aspects of the discovery will be instructive.

According to Kedrov (1966), Mendeleev's first "flash" consisted in the "idea of comparing chemically dissimilar elements in terms of the magnitudes of their atomic weights" (p. 21). As mentioned earlier, this flash can be traced back to both Mendeleev's conception of unity, which included relationship between dissimilar elements (unlike the opposing conception of unity), and the notion of an all-embracing generality. Comparing dissimilar groups in terms of their atomic weights is only warranted if it is theoretically implied that atomic weights belong to elements and that there is a relationship between dissimilar elements. Both assumptions can be seen as Mendeleev's original abductive hypotheses. What is most interesting is that "the groups first compared did not bring to light any regularity in the differences in the atomic weights of their members" (p. 21). Still, for

Kedrov, this was "the general key to the solution of the whole problem" (p. 21). What remained was the "exceptionally difficult" and "highly laborious" task of testing out the enormous number of possible combinations out of which it was hypothesized that a single general law will be revealed (p. 21). The entire process of Mendeleev's discovery, from his initial flash to the general law can be summarized, once again, with the help of Charles Sanders Peirce, and particularly, his notion of three distinct logical inferences.

According to Peirce, inquiry begins with doubt and surprise, in response to which an abductive hypothesis is generated about the nature of that which is causing the doubt (X) which in turn serves to overcome what is surprising about X (Gilmore, 2003). Understanding the cause of doubt is to understand the nature of X, "…to understand the general principles and relations that characterize it, which is to overcome its otherness and transform it into yet another element of the world that one understands…" (p. 388). Furthermore, even though abduction is a logical inference for Peirce, it has the nature of a guess, where the abductive inference is a guess directed toward the nature of X. Mendeleev's doubt was the question of the relationship between the elements. The abductive hypotheses of Mendeleev consisted in assuming both that atomic weights belong to elements *and* his particular conception of unity which included regularity in dissimilarities. These assumptions made what Kedrov calls Mendeleev's "key insight" possible, that is, that chemically dissimilar groups of chemically similar elements are periodically related by virtue of their atomic weights—Mendeleev's periodic law as X. However, the abductive insight serves only as a point of departure toward a more formal and clear understanding of X. In other words, the "exceptionally difficult" and "highly laborious" tasks of deduction and induction still remained. In order to inductively test the abductive hypothesis, definite and testable deductive corollaries of the abduction had to be iterated. Kedrov (1966) has delineated the specifics of Mendeleev's deductive iterations. What is important in the context of this paper is the efficiency with which Mendeleev performed this "exceptionally difficult" and "highly laborious" task—the second phase of his discovery—completing it in only one single day. The question then is: What accounts for the *efficiency* of this second stage?

The Question of Mendeleev's Efficiency

Psychological Factors

As Mendeleev had discovered the "general key" on the same day he had to leave to inspect the artels, he was under severe temporal limitations, or what Kedrov (1966) calls "Zeitnot." These limitations serve to show that Mendeleev could not go through all possible or even probable lines of inquiry, which would have taken several weeks, and was forced to make choices—eliminating potentially fruitful lines of inquiry. What accounts for Mendeleev's ability to guess right under such "unfavorable conditions?" Kedrov (1966 p. 20) suggests that it was precisely these limitations which focused all his will and creative energies toward answering a single question, albeit involving numerous steps. This goal directedness, leading to the strictly selective and conscious work of attention, magnified Mendeleev's ability to identify very definite signs pointing toward a hitherto hidden generality. In Kedrov's words, Mendeleev's search for these "definite signs" was guided by an already "sensed lawfulness" which consisted in identifying and eliminating "from among the numerous relations among elements—including the relation among their atomic weights—that which would provide a unified arrangement of the groups of elements in a definite order" (p. 22). Yet, this process of identification and elimination required keeping a systematic record of the known relationships while simultaneously envisioning novel relationships. This involved an unimaginable interplay between the powers of memory and imagination; this is what characterizes the second stage of the discovery for Kedrov.

Semiotic Considerations

The psychological phenomena mentioned above were undoubtedly necessary, yet are in no way sufficient to account for the efficiency of the discovery: "If one takes seriously the idea that a study of theorising has to concentrate on the concrete practices of scientists…one has to focus on the cognitive interactions between agents and the representational devices they reason with and manipulate" (Vorms, 2011, as cited in Campbell, 2017, p. 199). Insofar as Mendeleev's efficiency is related to his "cognitive interactions" with the "representational devices" he reasoned with, a deeper look at the cognitive tools—facilitating his memory, will, and creativity—is warranted. Kedrov (1966) tells us that Mendeleev was fond of the card game "patience" in which cards are arranged systematically, according to their suit and number. The resemblance between Mendeleev's predicament and the game of patience is uncanny: Both required a systematic arrangement of their respective elements under the guidance of two primary variables—atomic weight corresponding to card number and chemical properties corresponding to suit, arranged in rows and columns, the way Mendeleev arranged his periodic table. Mendeleev's history of playing patience not only honed his skills in identifying systematic relations in tabular form, but, according to Kedrov, Mendeleev actually played a game of "chemical patience" in which he made paper cards of the known elements mentioning both the relevant variables (p. 26). Even though Kedrov's account of Mendeleev's representational device is instructive, Campbell (2017) gives a more general account of Mendeleev's periodic table as the Peircean icon, general enough to account for even the iconic character of chemical patience. A closer look at Campbell's argument will reveal not only how intimately the structure of the periodic table was related to Mendeleev's thought process, but also how it facilitated that process, further accounting for its efficiency.

The Periodic Table as an Icon

Peirce is perhaps best known for his logic as semiotics, identifying and classifying myriad distinct signs. According to Campbell (2019) the Peircean *icon* is the most relevant and appropriate sign which characterizes Mendeleev's periodic table. In fact, the entire process of Mendeleev's discovery, especially the second stage, can be seen as the process of constructing an icon. Campbell (2017) identifies three key features of the Peircean icon. I will present these in a different order than Campbell (for the purposes of this paper, such reordering fits more neatly with Peirce's three logical inferences). The first characteristic of an icon is that it is a "system of relations," that is, it is not a mere thing but has a relational character (p. 228). What does this system of relations consist of? It consists of the relationship between chemical elements, their atomic weights, and their chemical properties. In Peircean language, the elements are symbols, the weights and properties are indices (singular index), and the icon represents their relationships. More technically, as the elements are arranged in groups and periods, it can be claimed that the periodic table is the iconic representation of indices drawing attention to families of symbols. However, the main point is that the icon is neither a mere thing nor monolithic, but rather a system of relations. This order is apparent to anyone who has ever seen the periodic table.

The second feature of an icon for Campbell (2017) is that it is both grounded in and supports "surrogative reasoning"—reasoning about an object based on a representation of that object aimed at disclosing new information about the object (p. 228). For Mendeleev, the object (of inquiry) was the real periodic law of nature he aspired to discover and it was represented via the periodic table. In other words, the icon is not merely a set of an arbitrary system of relations but a set of relations representing the relations characterizing a real object. It is by virtue of this similarity and affinity between the representational device and the real object that allows the inquirer to predict, or make explicit, that which is implicit in the real object by surrogative reasoning on the representation. This brings us to the third characteristic of the Peircean icon—"the possibility of making new discoveries about an object by observing and experimenting on its iconic representation"—its "epistemic fruitfulness" (Campbell, 2017, p. 228). The archival material presented by Kedrov (1966) shows that indeed Mendeleev was engaging in paper and pencil thought experiments on the day of the discovery. The icon, as a semiotic tool, facilitated Mendeleev's deductive thought process in making explicit the relations that were implicit in its construction. In other words, the icon made certain hidden relations perspicacious not by laboratory experimentation on the elements but via diagrammatic experimentation on the table itself. What were these hidden relations? They manifested as Mendeleev's prediction of unknown elements and his correction of the atomic weights of some known elements, not by a long and arduous process in the field, but via a highly efficient tool on paper. This, however, does not mean that Mendeleev engaged in pure speculation, as we have seen his distaste for it. Rather, the raw material (atomic weights and chemical properties) for his paper experiments was gathered inductively in laboratories by chemists; his work therefore consisted in coherently organizing the gathered data and filling in the gaps via predictions. Furthermore, Mendeleev's deductive predictions consisted not only of those previously mentioned, but also in the several iterations of the periodic table found in his archives. These deductive iterations were immediately tested by data that was already gathered inductively (the known elements) on the day of the discovery. The deductions corroborated in light of this data were the basis for Mendeleev's inductive generalization—the periodic law as represented by the periodic table. It is instructive to note here that Peirce, a trained logician, considered Mendeleev's inductive reasoning, from parts (atomic weights and chemical properties) to whole (the periodic law), as "one of the most admirable generalizations that the whole history of science can boast" (Peirce, 1900, as cited in Campbell, 2017). Peirce went as far as to claim that "very few inductions in the whole history of science are worthy of being compared with this as efforts of reason" (Peirce, 1892, as cited in Campbell, 2017, p. 14).

We are now in a position to better appreciate how Mendeleev's thought process was facilitated by the iconic character of the periodic table. His abductive hypothesis about the nature of the periodic law was facilitated by "surrogative reasoning"; his deductive predictions about unknown elements and the correction of atomic weights was facilitated by the icons "epistemic fruitfulness"; and his inductive generalizations about the periodic law were facilitated by the icons ability to hold a system of relations. However, it is very important to note that all three of Mendeleev's inferences and the accompanying three characteristics of the icon did not happen only once, but rather it is the continuous dynamic between all three which is of prime importance. This brings us to yet another feature of Mendeleev's periodic table as icon—"iconic robustness" (p. 163). To use the terms introduced previously, iconic robustness consists of the icon's continuing capability, as a system of relations, for surrogative reasoning and epistemic fruitfulness in light of what Campbell calls "nature's resistance" (p. 163). The resistance of nature consists of deductive predictions being falsified, or challenging the system of relations, revealing a flaw in surrogative reasoning—where the resistance is by virtue of the real object that the icon is representing. Put simply, the robustness of the periodic table as icon lies, paradoxically, in its flexibility or plasticity—its amenability to be reconstructed in light of contradictory evidence without losing its identity as an icon. At a subtle level, this was present even on the day of Mendeleev's discovery, as his initial formulations of the table reveal several crossings (failed deductive iterations in light of the gathered inductive facts) whereafter the form of the table was repeatedly reconstructed to assimilate outliers. The iconic robustness of the periodic table reveals itself more prominently, when "…the taxing problem of accommodating the rare earth elements and the discovery of novel elements such as the noble gases" (p. 165) emerged. The details of these resistances are not important here, but what is important to note is that several years after the periodic table had been presented by Mendeleev, novel evidence resisted the presented form. However, even though Mendeleev originally denied the contradictory evidence, with the help of other scholars, he was able to assimilate the new evidence by reconstructing his table in a manner that neither compromised its iconicity, nor completely dissolved its earlier structure. If anything, this assimilation further evidences the iconicity of the table and made the relations within it more perspicuous. In Peircean terms, the original structure of the icon (firstness) was resisted by nature or the real object (secondness), but was plastic enough to be reconstructed without losing its iconic identity (thirdness). To summarize, Mendeleev's discovery of the periodic law was intimately related to his construction of the periodic table, which in turn was not only able to reveal implicit relations of the periodic law but was also able to assimilate its novel, discovered relations.

The power of the icon has been discussed in terms of its iconic robustness, however, before proceeding, an interesting fact about its predictive, deductive power in relation to the memory of its constructor—Mendeleev—will be illuminating. According to Kedrov (1966), the archival records from the day of Mendeleev's discovery have two peculiar instances relating to the iconicity of the periodic table and Mendeleev's memory. The first was with "…respect to forgetting several conventions of notation, and recalling these inaccurately at various stages in his work on the periodic system" (p. 28). As Mendeleev was able to identify and correct these false notations quite quickly, the iconic robustness of the periodic table can then be seen as not only facilitating the assimilation of novel evidence, but also in correcting apparently false evidence. However, the other peculiarity reported by Kedrov completely eclipses the first one. The second peculiarity also relates to Mendeleev's forgetting, but in a different way: During Mendeleev's second stage of the discovery, when he was testing various iterations of the periodic table, Kedrov claims that the archival records indicate that Mendeleev actually predicted three noble gasses of an inert group unknown and undiscovered at the time. What is especially staggering and paradoxical about this is that Mendeleev did not pursue this line of inquiry further and that several years later, when the noble gasses were discovered, he actively denied their discovery; both times, the "denial" was grounded in how radically such elements would affect the nature of chemistry and the structure of his periodic table. Ultimately, he did assimilate these elements in his periodic table, but Mendeleev:

> …spoke of these discoveries as completely unexpected and unforeseen, although a quarter-century earlier he himself came very near to prophesying them…All of this testifies to the fact that after a lapse of 25 years Mendeleev simply forgot completely that he himself in point of fact predicted—though not as definitely as he did for several other then still undiscovered elements—the existence of at least three new elements, members of the future inert group.
>
> *Kedrov, 1966, p. 29*

It is argued here that this peculiarity can, in significant proportion, be accounted for in reference to the deductive power or epistemic fruitfulness characteristic of the periodic table's iconicity. It is astonishing that Mendeleev could not completely fathom the deductive power of a semiotic tool or representational system he himself constructed.

Abduction and Aesthetic Sensibility

We have seen how the question of Mendeleev's efficiency on the day of discovery can be answered, in part, by viewing his periodic table as a Peircean icon. For Kedrov (1966), however, another pressing question about the day of the discovery remains. To give some context, Mendeleev's discovery has been infamously tied to a dream he had on the day of discovery, a dream in which Mendeleev claims that "…all the elements fell into place as required. Awakening, I immediately wrote it down on a piece of paper…" (p. 30) Mendeleev's dream has been used by many to espouse an *enchanted* view of his discovery, in which the entirety of the periodic table is attributed to this dream. On the other hand, many scholars have denied the veracity of the dream altogether, espousing a *purified* view. Kedrov does not deny the veracity of the dream, yet, he does not espouse the enchanted view either, as he states:

> …first, the creative thoughts of Mendeleev even in his dream could continue his activity in the direction previously pursued, leading to the completion and improvement of the representation he had found earlier. Second, one can surely not speak of the discovery of the periodic law during sleep, although the testimony of Inostrantzev would give grounds for such a conclusion if we accepted it completely on faith and did not subject it to critical examination on the basis of the newly discovered archive materials. Several authors have made this error.
>
> *Kedrov, 1966, p. 31*

In other words, instead of approaching the issue of Mendeleev's dream as an either/or (purified vs. enchanted), Kedrov looks at it as both. However, the orthodox conception of both precludes the possibility of any relationship between them. How then can Kedrov's position—accepting the facticity of the dream while looking at it in continuity of his conscious activity—be understood? Once again, Peirce's unique conception of abduction helps answer this question. For Peirce, abduction is, first and foremost, a logical inference and simultaneously the only inference responsible for the discovery of novel positive knowledge (de Waal, 2016). If abduction is not conceived of as a logical inference and only as an enchanted, mystical flash, there can be no continuity between the known and the unknown. Yet, claiming that abduction is a logical inference does not mean that it is reducible to deduction and induction—the purified view. The form and character of abduction is distinct from both the other kinds of inferences. Still, Kedrov's emphasis is in showing that

> the entire process of discovery shows that there was nothing in it that would argue for complete suddenness and unexpectedness of its origins, any incoherence in its course, or the presence of any inexplicable "leaps," surprises or "revelations from Heaven." On the contrary, the newly discovered material allows us to establish this process as a fully connected, strictly ordered growth of the ideas of the scientist, though accomplishing very quickly, in the course of a day, a very important work that under other conditions would have taken very much more time.
>
> *Kedrov, 1966, p. 32*

Although Kedrov was unaware of both Peirce and abduction, Peirce's abduction helps in making Kedrov's explication of Mendeleev's discovery process more comprehensible. Surely, Kedrov helps see the "fully connected, strictly ordered" process of Mendeleev's discovery; Peirce's abduction not only fully supports this idea but guards against interpreting Kedrov's assertion of continuity in a purely inductive and deductive manner (p. 32). To summarize, the debate surrounding Mendeleev's dream has been discussed; Peirce's conception of abduction has been presented as a tool for understanding not only Mendeleev's discovery process but also Kedrov's explication of it. The other pressing question mentioned above can now be dealt with more directly.

According to Kedrov (1966), "it is necessary to clarify what Mendeleev had in mind when he said that 'the elements fell into place as required.' But for this it is necessary at the outset to determine what table he was speaking about" (p. 30). Kedrov, through his archival analysis, identifies the table Mendeleev constructed after his dream but goes on to tell us something surprising about this table: He claims that, unlike the table constructed after the dream, the one constructed before the dream was apparently complete, but its arrangement did not exhibit the unbroken continuity of the series as a whole. The reason is that Mendeleev originally placed chemically similar elements (groups) in rows instead of columns (like the periodic table we see today) as he was interested in finding the differences in the atomic weights of chemically dissimilar elements, that is, it had arithmetic significance: This is the way we usually place numbers when finding differences— vertically and not horizontally. The problem was that, in this arrangement, atomic weights descended from left to right in rows and ascended from top to bottom in columns resulting in a discontinuity arising from the way the Russian language is written and read, from left to right—the last element in a row was discontinuous from the first element in the next row. In other words, even though the table was *logically* complete, there remained, what can be called an *aesthetic* discontinuity between the mathematical (top to bottom) and Russian (left to right) orientations. Interestingly, it was precisely this change in orientation, from horizontal to vertical, that was the revelation in Mendeleev's dream. Therefore, Mendeleev's yearning was not only to logically complete his periodic table, but to complete in a manner that represented "entirely successfully the whole pattern of the structure in such a way as to give at a glance the notion of the essence of the law" (Kedrov, 1966, p. 31). Again, Kedrov's analysis shows us that indeed, the dream was the next logical step in the process of discovery—once the differences were worked out, there was no reason to maintain the table's horizontal orientation at the cost of its aesthetics. Still, even though Kedrov shows us *what* Mendeleev meant by "the elements fell into place as required," he does not shed light on *why* Mendeleev was motivated to do so, especially considering that the table was already logically complete (p. 30). The personal accounts of the famous French mathematician, Henri Poincaré, will help shed further light on this unexplored, aesthetic dimension.

Poincaré's Insights

"The genesis of mathematical creation is a problem which should intensely interest the psychologist" (Poincaré, 2000). This is the first line of a short personal reflection by Poincare. Even though he is interested in the "genesis of mathematical creation," his call to the psychologist and the uncanny resemblance of his description of the mathematical creative process to Mendeleev's process of discovery warrants a deeper look at his *reflections* (p. 85). Beyond this, Poincare also helps answer the previous question by shedding light on the irreducible aesthetic motivation behind the creative process. Mathematical creation, is "guided by the general march of reasoning" and its fruit manifests itself in a mathematical demonstration (p. 87). This "mathematical demonstration" is much like the "form of presentation" Mendeleev was interested in, where this demonstration "...is not a simple juxtaposition of syllogisms, it is syllogisms *placed in a certain order*, and the order in which these elements are placed is much more important than the elements themselves" (Poincaré, 2000, p. 87). Even though it is the "certain order" here which is of primary interest, it is also interesting to note that Poincaré has used the word "elements," though in a different sense. For Poincaré, the process of creation involves a choice: distinguishing the minority of useful combinations from an overwhelming majority of useless, yet possible, combinations. Still, the peculiarity is that, for Poincaré, the useless combinations do not even enter the inquirers or creator's consciousness. This is glaringly obvious in Mendeleev's case as mentioned above; the conditions of Zeitnot forced him to make choices: Out of the large number of potential choices, the "right" one was revealed in the few iterations Mendeleev consciously worked out. Furthermore, for Poincaré, it is not the content, but the circumstances (like Zeitnot) that are of primary interest for the psychologist. Poincaré explains the general circumstances of the creative process involving three chief stages: First, there a period of intense, prolonged, and conscious activity directed toward solving a problem which appears to end in futility. Second, there is a period of unconscious activity, which is incontestable to Poincaré, manifesting itself in the form of a "sudden illumination, a manifest sign of long, unconscious prior work" (p. 90). What is characteristic of this sudden illumination is that it does not reveal the answer as a finished product; the illumination is akin to a vague feeling rather than a definite formulation, and thus, only serves as a point of departure for further conscious activity. This further conscious activity marks the third and final stage of the creative process, consisting in the disciplined, laborious, and willful deducing of consequences of the "intuition" and verification of results. Once again, the similarities with Kedrov's account of the day of discovery are uncanny: Mendeleev had a whole career of conscious activity behind him, the initial insight revealing the "general key," what Kedrov calls a "flash," was followed by the exceptionally difficult and highly laborious work. Yet again, Mendeleev's dream ["a manifest sign of long, unconscious prior work" (p. 90)] can also be seen as the central period preceded and followed by a laborious and willful conscious period. Furthermore, Mendeleev's sudden illumination can be seen as an abductive hypothesis, in logical continuity with previous conscious activity but requiring the further conscious activity of deduction and induction, which was made highly efficient by his use of an icon. Still, a question remains for Poincaré: Why do only a few useful combinations enter consciousness? This question is intimately related to the question above—what Mendeleev meant by "the elements fell into place as required." According to Poincaré (2000), the irreducible role of unconscious activity is incontestable. The reason that out of the potentially infinite combinations, the right one is revealed in the few combinations that enter our consciousness (abduction as guessing right), is for Poincaré, like Peirce, *not* attributable to chance. Rather, Poincaré hypothesizes that the reason that only these few possibilities draw our attention is attributable to the same reason any stimulus enters into our awareness:

> ...among all the stimuli of our senses, for example, only the most intense fix our attention, unless it has been drawn to them by other causes. More generally the privileged unconscious phenomena, those susceptible of becoming conscious, are those which, directly or indirectly affect most profoundly our emotional sensibility... This is a true esthetic feeling that all real mathematicians know, and surely it belongs to emotional sensibility.
>
> *Poincaré, 1910/2000, p. 92*

Poincaré, goes on to ask which "mathematical entities" (Peircean reals) are responsible for affecting, so profoundly, our emotional sensibilities:

> They are those whose elements are harmoniously disposed so that the mind without effort can embrace their totality while realizing the details. This harmony is at once a satisfaction of our esthetic needs and an aid to the

mind, sustaining and guiding. And at the same time, in putting under our eyes a well-ordered whole, it makes us foresee a mathematical law.

p. 92

In other words, the goal-directed conscious activity of the inquirer toward truth is complemented by the unconscious emotional sensitivity toward beauty. This warrants the conclusion that "…the useful combinations are precisely the most beautiful" (Poincaré, 1910/2000, p. 92). This is not to say that all truth is conscious and emotional sensibility is unconscious, that would preclude any relationship between them, but rather to highlight what is most characteristic of both. Beauty is then an objective quality for Poincaré, revealed to the inquirer's emotional sensitivity, fixing the inquirer's gaze and guiding the inquirer's mind toward its expression in the form of a law. This brings us to the infamous Roycean triad of beauty, duty, truth (Oppenheim, 1993). The inquirer considers it a duty or responsibility to express the beautiful in a lawful manner. It then makes sense why aesthetic considerations, pertaining to the form of the product discovered or created, are equally important for the inquirer as the *real entity* discovered is not only perceived to be true—it is also perceived to be beautiful. Therefore, the inquirer yearns to bring out not only its lawful character, but also its beauty (harmony)—in fact, this is what characterizes the inquirer's "duty." It is argued here that Poincaré's hypothesis coherently explains the meaning of Mendeleev's statement. Mendeleev's yearning to represent or demonstrate the "…whole pattern of the structure in such a way as to give at a glance the notion of the essence of the law…," even after his table was logically complete, echoes Poincaré's characterization of beauty as that "…whose elements are harmoniously disposed so that the mind without effort can embrace their totality while realizing the details" (Kedrov, 1966, p. 31; Poincaré, 2000, p. 92) The final form on which Mendeleev was content was then one which simultaneously satisfied his aesthetic sensibilities and logical faculties, transpiring through a continual struggle with the empirical data.

Another Peculiarity

Poincaré's hypothesis also helps explain another peculiarity reported by Kedrov (1966). For Kedrov, on the day of the discovery, the sign which betrays that a flash had revealed the "general key" to Mendeleev, consists in Mendeleev's comparison of groups in terms of their atomic weights, as evident in the archival records of that day. However, the peculiarity is that the initial groups compared did not reveal any regularity; if inductive evidence is the only reason to take a hypothesis seriously, what explains Mendeleev's tenacity in holding on to a falsified hypothesis, trying to "prove" it through another iteration? It is argued here that Poincaré's hypothesis (satisfaction of an emotional or aesthetic sensibility) serves to explain this peculiarity and fits well with Gerald Holton's (2000) notion of "suspension of disbelief"

(p. 96). Holton argues that contrary to the "purified" textbook view of how science proceeds, the history of science reveals a counterintuitive element in the trajectory of discovery. This element manifests itself in the conduct of scientists—tenaciously holding on to aesthetically determined thematic commitments, not only in the absence of positive evidence, but also in the face of contradictory evidence. The discoveries of great scientists like Copernicus, Galileo, and Einstein reveal irreducible thematic commitments at play. Holton, claims that "Copernicus is a case study of the privileging of an aesthetically based theory," speaks of Galileo's "…enchantment with the circle," and states that Einstein's theory was regarded by his early followers (scientists) as having "…the highest degree of aesthetic merit; every lover of the beautiful must wish it to be true" (2000, p. 60). Even though Mendeleev's "suspension of disbelief" is not as pronounced in the peculiarity quoted above, his lifelong unease with the orthodox notions of atom, substance, and unity; his yearning to represent the periodic table in harmonious form; the fact that such thematic commitments have been part of major scientific discoveries in the past; all serve as warrants to take the aesthetic and thematic dimensions, involved in the process of his discovery, seriously. In sum, the role of the "emotional sensibility" in mathematical creation alluded to by Poincaré, and the role of the "thematic imagination" suggested by Holton are not only consistent with each other, they also serve to explain Mendeleev's lifelong unease with commitments to a different kind (a particular kind of unity and hence beauty), his peculiar suspension of disbelief, and his yearning to represent the periodic table in a beautiful form.

Mendeleev's Personality

Mendeleev's process and context of discovery have been discussed above. Also, an attempt has been made to answer various questions that arise in relation to his discovery. However, Kaji (2018), at the end of his chapter on Mendeleev, asks a question of a different kind. In relation to another chemist working on the periodic law, Kaji asks "why was Mendeleev so bold?" (p. 238). Before proceeding toward an answer, this question may be framed as a more general question about Mendeleev's personality: What hints do we find about his dispositions and traits? First of all, we have already discussed the pragmaticist trajectory of Mendeleev, this characterizes his intellectual style. What is the pragmaticist trajectory but a focus on sensible effects and how they relate in a coherent whole? This may be called a pragmaticist or gestalt awareness. The life of Mendeleev is rife with examples of such gestalt awareness characterized by his ability to see things in related wholes— the assumptions behind and futural implication of ideas. Examples include his awareness of the relationship between his national, professional, and pedagogical activities; appreciating the implications of indefinite compounds for atomic theory; the philosophical and religious implications of his periodic law (Bensaude-Vincent, 1986); and even the trajectory of his research in his late life, searching for a "true physical foundation of the periodic law," (Kaji, 2018 p. 236) reveal his gestalt awareness. Second, coming to Kaji's question, the second cardinal feature characterizing Mendeleev's personality was his audacity and boldness; particularly, his audacity in asserting his gestalt awareness. Tenaciously challenging the mainstream or orthodoxy at multiple fronts on the basis of novel conceptions generated by his own gestalt awareness is a prime example. Also, predicting new elements—something that was completely alien to the work of chemists at the time—with such accuracy, going as far as to assert that the experimental data were wrong and that the experiments should be performed again, is also a manifest sign of his boldness. Kaji (2018) attributes this boldness to Mendeleev's early success in chemistry which gave him the confidence needed to pursue "neglected" lines of inquiry and "the support and encouragement of the Russian chemical community, as well as German mediation" (p. 239). Finally, the third and seemingly paradoxical character trait of Mendeleev, as revealed by his personal letters, is humility. At first, it seems counterintuitive to simultaneously assert the apparently opposite attributes of audacity and humility. A closer look, however, reveals this to be the case. After publishing his initial periodic table in 1869, Mendeleev received a letter the same year by Nikolaevich Zinin, the first president of the Russian Chemical Society, criticizing his nonexperimental work and urging him to do "real work" (Kaji, 2018). It appears that experimental, not theoretical, work was required (implicitly) by a chemist to be identified as a true scientist. This suggests that a certain amount of social pressure was involved in pursuing a theoretical line of inquiry—another sign of Mendeleev's tenacity. However, what is important to note about this letter, which Mendeleev ultimately did not send, is the relevance of its tone and content to the present discussion. Mendeleev wrote:

> Even if Germans do not know my works, it is understandable and does not annoy me. I will not take any measure to make Germans know my works. When what I had done was stolen by others…I would not say a word, because I do not have self-deception which is flagrant and harmful for science and because I despise priority disputes.
>
> *Kaji, 2018, p. 233*

Mendeleev goes on to tell Zinin to evaluate the content of his work on its merit, and not to tell him what kind of work he should concern himself with. What is of greater interest here is Mendeleev's lack of concern about being the center of attention in relation to his work and contributions. It may be that Mendeleev's audacity was in relation to using his gestalt awareness and pragmaticist trajectory toward the progress of science and Russia. However, using his personal accomplishments in order to celebrate his person was clearly not his priority. In sum, Mendeleev's simultaneous audacity and humility, mediated by his gestalt awareness, characterize his personality.

Relevance and Questions

What is the significance of the previous discussion? How is it relevant to contemporary issues? What further questions are generated through such an exploration? Burman (2018) claims that "the history of psychology has two aspects: the content and the activity. The stuff and the doing" (p. 1). What is said about the history of psychology also holds true for the history of scientific discoveries. These discoveries are usually bifurcated into the logical content of the discovery and the historical form taken by that discovery. The multifaceted history of Mendeleev's discovery reveals that his actual historical context cannot be divorced from the content he discovered. A deeper look into the context of this discovery has revealed how his ultimate discovery was contingent upon a local and concrete sociological, theoretical, and personal context. This paper ends with the hypothesis that this concrete and local context is a feature of the discovery itself and paying attention to it, that is, appreciating its deeper relations with the content of the discovery will bring to light the general conditions under which the discovery process is facilitated. These general conditions would be especially valuable to pedagogy. Based on the discussion in this text, a few examples follow:

> The use of the icon was cardinal to the final stages of Mendeleev's discovery. As such, the Peircean Icon is a cognitive device and a semiotic tool through which the inquirer reasons. Students are usually taught to read and write in order to develop their critical thinking. However, the use and significance of semiotic devices such as diagrams is hardly ever a central pedagogical concern. Mendeleev was a chemist, still, anyone who has taken a chemistry class knows the importance of diagrams and representations. The question is, how relevant is the art of diagramming relevant to scientific inquiry? Teaching students how to use and work with advanced cognitive representational tools holds promise to develop more efficiency and fecundity in their thought.

Mendeleev was able to see which empirical facts were related to his philosophical assumptions. When contradictory facts were brought to his attention, he was able to appreciate how such facts undermined the credibility of assumptions grounded in a different set of facts. This allowed him to explore possibilities outside the parameters set by his original assumptions without which he could not have discovered the periodic law. Students today are usually not familiar with the sets of assumptions they are working with, let alone philosophy of science in general. Teaching students about the limitations and assumptions of their respective paradigms is not only important for them to have a more informed view of where the stand, but also a prerequisite to explore alternative sets of assumptions if contradictory evidence emerges.

The history of science in general and Mendeleev's history in particular reveal that major scientific breakthroughs are often not based on "hard evidence" (inductive data) but aesthetic preferences (Holton, 2000). What is the relationship between *opening our eyes to facts* and *opening our eyes to beauty*? Are they mutually exclusive, or is there a more intimate relationship between these distinct *visions*? The discussion above only hints at the relationship between the aesthetic and the scientific. However, it warrants more critical attention toward the nature of this relationship. Such an exploration holds the promise to reveal hitherto unknown affinities between art and science, affinities which would have serious and far-reaching implications for pedagogy.

There are several other pertinent questions which could be generated. For example, what personal characteristics or "virtues" were important for Mendeleev's scientific discoveries, and what might be the broader implications for science? What roles can dreams and other unconscious processes play in scientific discovery? Is the inquiring scientist insightful and competent enough to explicate the process of their own discovery? If not, which others are to be granted such authority? If both the scientist's and others' interpretations are to be taken seriously, how are these, possibly contradictory, interpretations to be negotiated? These are only a few examples of how inquiring into the process of inquiry holds the promise of opening further inquiry.

Conclusion

Textbook accounts of scientific discovery tend to emphasize the *universal* content of those discoveries without due consideration to the *local* activity of discovery. A closer look at the activity of discovery reveals that it is embodied by a unique agent implicated in complex social networks and historical trajectories. Without due consideration to these intricate relations, any account of scientific discovery is wanting. Moreover, a critical examination of these relations promises not only a more informed view of science, but invaluable pedagogical lessons conducive to the forward movement of scientific activity. The various elements considered in this paper include Mendeleev's patriotic and scientific motivations, intellectual zeitgeist, conceptual development, philosophical commitments, psychological processes, representational devices, dream, aesthetic ideals, intellectual style, and dispositions. To use the language and metaphor of chemistry, all these elements have their unique weightage toward Mendeleev's discovery by virtue of their effects (properties) on his trajectory. Many of these elements can be grouped together—such as his focus on a particular kind of unity for atom, element, and law—and a general relationship (law) can be discovered between the various periods of his life by virtue of these grouped elements. This generality is much like the generality Mendeleev himself was interested in—one that does not compromise on individuality. The general relationship holding all these elements together is Mendeleev the person.

QUESTIONS FOR DISCUSSION

1. What is the relationship between opening our eyes to facts and opening our eyes to beauty? Are they mutually exclusive, or is there a more intimate relationship between these distinct "visions"?
2. What personal characteristics or "virtues" are important for Mendeleev's scientific discoveries, and what might be the broader implications for science?
3. What roles can dreams and other unconscious processes play in scientific discovery?
4. In what important ways do "signs" and representations contribute to discovery and demonstration?

References

Bensaude-Vincent, B. (1986). Mendeleev's periodic system of chemical elements. *The British Journal for the History of Science, 19*(1), 3–17.

Burman, J. T. (2018). What is history of psychology? Network analysis of Journal Citation Reports, 2009–2015. *SAGE Open, 8*(1), 2158244018763005.

Campbell, C. (2019). The periodic table as an icon: A perspective from the philosophy of Charles Sanders Peirce. *Centaurus, 61*(4), 311–328.

Campbell, C. J. (2017). *The chemistry of relations: The Periodic Table examined through the lens of CS Peirce's philosophy* (Doctoral dissertation, University College London.

de Waal, C. (2016). "Charles Sanders Peirce and the abduction of Einstein: On the comprehensibility of the world." *arXiv preprint arXiv:1610.00132.*

Dewey, J. (1920). *Reconstruction in philosophy.* Henry Holt and Company.

Gilmore, R. (2003). Peirce and perversity: The higher logic of the real. *Transactions of the Charles S. Peirce Society, 39*(3), 383–404.

Holton, G. J. (2000). *Einstein, history, and other passions: The rebellion against science at the end of the twentieth century.* Cambridge, MA: Harvard University Press.

Kaji, M. (2018). The origin of Mendeleev's discovery of the Periodic Table. In E. Scerri & G. Restrepo (Eds.), *Mendeleev to Oganesson: A multidisciplinary perspective on the Periodic Table* (pp. 219–244). Oxford: Oxford University Press.

Kedrov, B. M. (1966). On the question of the psychology of scientific creativity. *Soviet Psychology, 5*(2), 18–37.

Oppenheim, F. M. (1993). *Royce's mature ethics.* Notre Dame, IN: University of Notre Dame Press.

Peirce, C. S. (1878). 1958. How to make our ideas clear. *Charles S. Peirce: Selected Writings,* 113–141.

Peirce, C. S. (1974). *Collected papers of Charles Sanders Peirce (Vol. 2).* Cambridge, MA: Harvard University Press.

Poincaré, H. (2000). Mathematical creation. *Resonance, 5*(2), 85–94.

Slife, B. D., Williams, R. N., (1995). *What's behind the research?: Discovering hidden assumptions in the behavioral sciences.* Thousand Oaks, CA: Sage.

4

HENRI POINCARÉ

The Poet of Mathematics and Physics

Yousaf Raza

Personal Preamble

My choice to write on Henri Poincaré was straightforward. At the time, it had little to do with him and a lot to do with my allegiance to my culture. As a Pakistani psychiatrist, I was acutely aware of the difficulty that psychiatry faced in meaningfully relating to Pakistani culture. I came to understand, studying Charles Peirce with my teacher Dr. Basit Koshul, that this difficulty is rooted in the philosophy of science that underpins psychiatry. In other words, it was due to hidden assumptions regarding the scientific method that psychiatry adhered to. I was taught that Poincaré explored these hidden assumptions. As a mathematician and physicist, he explored the philosophy and psychology of his own scientific process. In doing so, Poincaré expands the understanding of the scientific method in a way that makes it far more Peircian and far more relational. In order to truly live up to its ends in Pakistan, psychiatry needs to repair its understanding of philosophy that underpins clinical practice along Peircian lines, corroborated by Poincaré. This improvement of psychiatry's relationship with its own philosophy will render it more relational and as a result, it can genuinely relate with Pakistani culture.

Hans Eysenck (1995) describes the mathematician Srinivasa Ramanujan as the "Indian intuitionist *par excellence*" and Godfrey Hardy, his mentor at Cambridge University, as the "rigorous analyst" (p.195). Eysenck's psychological exploration of the nature of genius leads him to place them on either end of an intuition–logic continuum (Eysenck, 1995, p.186). It took Hardy's disciplined analytical rigor to tutor the "entirely original and incredibly creative" Ramanujan to generate the necessary proofs for his intuitions (Eysenck, 1995, p.196). It took Eysenck's background in experimental psychology to describe the relationship between intuition/Ramanujan and logic/Hardy as such.

DOI: 10.4324/9781003276692-6

Stern (1938) utilized the phrase *"unitas multiplex"* to portray the multiplicity held in consonance within a person (p. 73). In Henri Poincaré, we see some dimensions of this multiplicity refined to a remarkable degree. He brings together within himself the intuitive genius, the analytical mathematician, and scientist. In presenting his reflections on his own creative and analytical process, he performs the function of a psychologist of science. Scarcely seen in one person, these dimensions of Poincaré's personality and work mean that he is uniquely placed to buttress a pragmatic philosophy of science in place of empiricist modernism, or a "deconstructive postmodernism" (Griffin, 2000, p. x). In detailing how Poincaré can make these diverse, yet deeply intertwined, contributions, I explore the person that he was.

Henri Poincaré—The Person

Jules Henry Poincaré was born on 29 April 1854, to Leon Poincaré and Eugenie Lauonis. Leon Poincaré was a medical doctor serving in various capacities as a surgeon, neurologist, physician, teacher, and researcher working for the government, privately, and for charity medical services. He was an "inflexible worker" and, as his family feared, overworked himself to death. After suffering a head injury, he didn't take the prescribed rest and succumbed to his injury. He was eulogized at his funeral as a "sagacious and penetrating mind" and a "generous heart" (Ginoux & Gérini, 2014, pp. 6–9).

His father was a workaholic, and that naturally meant that Poincaré saw very little of him. The significant interactions described between father and son are mainly work-related, assisting his father in providing medical relief during the Franco-Prussian war in 1870 (Weinstein, 2012, p. 10) and assisting him in arranging a congress on the progress of science in 1886 (Verlaust, 2012, p.64).

The impact that his father had on him is easily mapped out. To the extent that intelligence and conscientiousness can be attributed to a genetic endowment, it is clear that H. Poincaré was copiously endowed. The degree to which these two attributes are nurtured and inspired can also be noted in the father–son interactions. Conscientiousness, whether understood in terms of industriousness or moral responsibleness, can be seen in both individuals. Indeed, Poincaré's advocacy of "Science as Value," as the title of one of his books and manifested in his own life, can be traced to his father's dedication to science.

Not surprisingly, we find little detail about Eugenie in Poincaré's biographies (at least those available in English). Nevertheless, the little we find helps us understand her significance in Poincaré's life and his achievements. Verhulst (2012) describes her as intelligent and lively (p. 3) and as having given him "plenty of loving attention" to make up for his father's absence (p. 5). She is a constant presence in his life; as the addressee whenever a teacher identifies him as brilliant in mathematics, as the one taking precautions against his "absent-mindedness" in aligning his pocketbook with bells (Ginoux & Gérini, 2014, p. 14, 44), as the witness to his obsessing over a mathematical problem wandering all around the house, and as the correspondent for hundreds of letters whenever he would live away from her (Verhulst, 2012, p. 15, 19). Indeed, the most distressing time in Poincaré's life was when he had to deal with the grief of losing his mother on 15th July 1897. Verhulst (2012) describes the grief period thus:

> Henri was shattered by this event, and for several months thereafter, he did not answer any letters and kept himself confined to his family. In a note to the Swedish mathematician Gösta Mittag-Leffler [Poincaré, 1999, letter 143], he wrote on July 31 of that year that he was unable to work and that he could not discuss any requests for refereeing or editing papers. On August 3, Mittag-Leffler replied with understanding and sympathetic condolences. It took until October 11 for Poincaré to resume his correspondence.
>
> *p. 45*

The second significant source of stability for Poincaré, and his most trusted confidant, was his sister, Aline Poincaré. She was two years younger than him and played the most significant role in rescuing the genius of her brother when she was only three years old. At five years of age, Poincaré was diagnosed by his father with Diphtheria. He is said to have shown signs of intelligence early on, attaining all his developmental milestones, especially speech, far earlier than most children. However, his illness took away his speech and left him paraplegic. The infection threatened to be fatal and had him spend the contagious period of his illness in isolation from his sister. He survived the disease, but it left him paraplegic and unable to speak. He recovered mobility early (taking his first step holding Aline's little hand), but his speech did not return for another nine months. The severity of the illness and its duration were sufficient to have left him with some form of learning disability. However, Poincaré's creativity found expression in developing a personal sign language through which he communicated with his sister. It is hypothesized that this not only helped him preserve his intellectual prowess, it helped him develop a facet to his creativity that was to later manifest in the uniqueness of his genius. It left him awkward in gait and shy in speech (Verhulst, 2012, pp. 8–9). He would avoid outdoor sports and spend more time indoors inventing games for his sister and cousins (Weinstein, 2012, p. 7).

Poincaré's genius expressed itself once he started school. He was top of his class until seventh grade. And for the rest of his time as a student, he was at least in the top three. He was labeled a "mathematics monster" (Verhulst, 2012, p. 9) while exhibiting a considerable brilliance in literature (Ginoux & Gérini, 2014, p. 13) and keen interest in science as well (Weinstein, 2012, p. 9). The exceptional range of ability and interest made it considerably difficult for him to decide what line to pursue as he confided to his sister, "I cannot commit myself to anything. I don't know what I will do in twenty years' time" (Verhulst, 2012, p. 11). He went on to show considerable interest in engineering, physics, arts, philosophy, and even writing for stage (p. 13).

His way of studying was interesting, and for some people, annoying. He wouldn't sit at a desk and work through a problem step by step on a standard notebook with a dominant hand. Instead, he would be seen walking around, engaging in conversations even, and then returning to the problem he was working at, and without bothering to sit, sometimes with his left hand and sometimes with his right, jot down the solution (Ginoux & Gérini, 2014, p. 13).

The Franco-Prussian War didn't just lead Poincaré to perform a social duty in assisting his father; it also led him to learn the German language sufficiently to understand the news regarding the war (Verhulst, 2012, p. 12). He would refine this ability enough to correspond with Lazarus Fuchs (p. 36) and present papers in the German language (p. 60). Prussian forces occupied Nancy and the French lost the war and many cities to the Prussian empire. Did the war impact his intellectual development in any negative way? The war delayed his exams, and once he was able to complete the exam finally, he scored a zero in mathematics. Curiously, the temporal relationship between the war and his zero in the exam is clear from the biographies. Still, the failure is attributed to his absentmindedness more than anything else.

His exceptional past record and overall performance in the science exam allowed him to proceed to the Special Mathematics class, which he passed with an Honorary Prize at the Concours General. He went on from there to Ecole Polytechnique as Jules Henri Poincaré. Originally Henry, the secretary at the school, misspelled his name as Henri. Poincaré couldn't care less and continued as Henri for the rest of his life (Ginoux & Gérini, 2014, pp. 14–16).

At the Ecole, Poincaré wasn't sufficiently challenged. He demurred the military rigor and even stood his ground on mathematical problems in opposition to a teacher, too, if he had to. This resulted in losing the first place. He still retained second place, and also published his first article in Mathematics around that time. He proceeded to the Ecole des Mines, from where he was posted as "Inspector of Mines in charge of the mineralogical sub-district of Vesoul" (Ginoux & Gérini, 2014, pp. 17–23).

The short stint of eight months as a mining engineer left a lifelong impression on Poincaré. This is reflected in an article he was asked to write for children, encouraging them toward science. Poincaré cites his experience in the mines and how the invention of the Davy lamp has been instrumental in making mining a far safer excursion. In his writing, he alludes to the explosions in the mines during his stay there and how lives were lost, leaving widows and children. His emotional expression alludes to the traumatic nature of those experiences and how personal the value of science was for Poincaré. Also worth noting is how meticulously Poincaré performed the most mundane of his duties (Ginoux & Gérini, 2014, p. 27–31).

With this begins Poincaré's professional career as a scientist. Before moving on to his achievements as a scientist, it is worthwhile to note Poincaré's attitude toward religion. Poincaré took his sacrament at age eleven very seriously, according to his sister (Verhuslt, 2012, p. 10). At eighteen years of age, he pronounced himself a free thinker. He felt Catholicism impeded the right to search and tell the truth (Weinstein, 2012, p. 12). However, his thought reflects a more nuanced stance in one of his lectures toward the end of his life. Poincaré channels his criticism against dogma which, he states, may afflict a scientist regardless of religiosity. He also appreciates the indispensability of the services of the likes of Pasteur, who regardless of their religiosity make invaluable contributions to science (Verhulst, 2012, p. 60). He had a good personal and intellectual relationship with Aline's husband, Emile Boutroux (Poincaré's brother-in-law). Boutroux was a philosopher known for defending the compatibility of religion and science, and one of his students was the eminent Henri Bergson (Verhulst, 2012, p. 236). Poincaré alludes to the implications of his work explicated by Boutroux in *Science et Religion* (Poincaré, 1913, p. 183*)*.

Henri Poincaré—The Scientist

Poincaré's contributions led to various developments in both mathematics and physics. The discovery of the Fuchsian and Kleinian Functions are the most significant among those contributions. They were a result of his rigorous efforts, detailed correspondence with Lazarus Fuchs and Felix Klein, and as he would go on to describe in *Science and Method* as "unconscious" and "intuitive" work (Ginoux & Gérini, 2014, pp. 35–39). As much as his discovery of the Fuchsian functions speaks to a victory of French academia over their German counterparts (p. 37), his naming them "Fuchsian," earning the displeasure and criticism of Klein and acknowledging the work of the German mathematician, speaks of Poincaré's determination to remain judicious.

His discovery of the "Terra Incognita," instrumental in developments within differential equations, led him to earn the prize of King Oscar II of Sweden and Norway in 1889 (Ginoux & Gérini, 2014, pp. 45–48).

Verhulst (2012) credits Poincaré's versatility for creating new research fields including, but not limited to, "automorphic functions, uniformization, the qualitative theory of differential equations, bifurcation theory, asymptotic expansions, normal forms, dynamical systems, integrability, mathematical physics, topology" (p. 99).

Poincaré's work was also instrumental, along with Lorentz, for Einstein's later development of special relativity. Verhulst (2012) quotes Lorrentz's appreciation of Poincaré's contributions, "Poincaré, on the other hand, has obtained a perfect invariance of the equations of electrodynamics and he has formulated the 'relativity postulate' in terms that he was the first to use," and goes on to express his perplexity at Einstein not crediting Poincaré in his 1949 address describing the development of relativity (p. 64).

It is worth noting that Poincaré's contributions concerning the theory of special relativity came at a time when Newtonian mechanics was, in Poincaré's own words, "strongly shaken" and the "new mechanics" was a work in progress (Verhulst, 2012, p. 208).

Henri Poincaré—The Psychologist

Poincaré alluded to the importance of psychology of science a hundred years before the field would develop a significance in the form of a subdiscipline. Indeed, Poincaré's recognition of the significance comes across in various ways in his biographical account; Verhulst (2012) notes that he was interested in graphology (p. 24) and that he presented himself for a psychological analysis to French Psychologist Eduard Toulouse (p. 98). Perhaps his most significant contribution to the field of psychology, as I will discuss toward the end of this section, is his description of the circumstances surrounding his discovery of the Fuchsian functions (Poincaré, 1913, pp. 179–185).

As for graphology, Poincaré's interest was sparked by his surprise as to how women have "neat and well-ordered handwriting" despite being very haphazard in thinking and somewhat lacking in logic (Verhulst, 2012, p. 24). Poincaré's penmanship was described as "very bad, and his draftsmanship even worse" (Ginoux & Gérini, 2014, p. 6). He went on to study graphology, analyzed his own handwriting, and concluded that he is impatient, pliable, and devoid of "bureaucratic feelings" (Verhulst, 2012, pp. 24–25). Toulouse's "not very exciting observations" are summarized by Verhulst (2012):

> 1. He worked during the same times each day for short periods. Mathematical research took four hours a day, two in the morning and two hours from 5 p.m. till 7 p.m. 2. His normal work habit was to solve a problem mentally and then write it down. 3. He was ambidextrous and near-sighted. 4. He could very well memorize and visualize what he read and heard. 5. He was physically clumsy and not artistically gifted. 6. He was always in a hurry and hated going back for corrections. 7. He believed in letting his unconscious work on a problem while he consciously worked on another problem.
>
> *p. 98*

In terms of biography and psychology, Poincaré's psychological analysis of his method of mathematical creation is far more incisive and significant. He appeals to the psychologist to take a keen interest in the subject, saying:

> The genesis of mathematical creation is a problem which should intensely interest the psychologist. It is the activity in which the human mind seems to take least from the outside world, in which it acts or seems to act only of

itself and on itself, so that in studying the procedure of geometric thought we may hope to reach what is most essential in man's mind.

Poincaré, 1913, p 179

Poincaré (1913) begins the discussion inquiring into why, despite its logical nature, mathematics isn't easily understood by most people and how the sanest and most logical of minds are prone to error in mathematics. He identifies two critical faculties for mathematical understanding: intuition and memory (pp. 179–180). He gives priority to intuition, stating that one can have a memory that isn't very exceptional, citing his own example, but intuition can help make up to facilitate mathematical understanding:

> If I have the feeling, the intuition, so to speak, of this order, so as to perceive at a glance the reasoning as a whole, I need no longer fear lest I forget one of the elements, for each of them will take its allotted place in the array, and that without any effort of memory on my part.
>
> *p. 180*

Intuition can help reinvent the process where memory naturally fails. Those gifted in terms of memory, he continues, but not so much in intuition can learn and sometimes apply, but not create. Creativity, he contends, emanates from being gifted in intuition (Poincaré, 1913, p. 180).

What does Poincaré mean by mathematical creativity? He specifically identifies this as the mathematician's ability to make the right choice when he says, "Invention is discernment, choice" (Poincaré, 1913, p. 181). What is it that is being chosen, and how is this choice made? Poincaré (1913) explains:

> ...the mathematical facts worthy of being studied are those which, by their analogy with other facts, are capable of leading us to the knowledge of a mathematical law just as experimental facts lead us to the knowledge of a physical law. They are those which reveal to us unsuspected kinship between other facts, long known, but wrongly believed to be strangers to one another.
>
> *p. 181*

Most of the possible combinations to be chosen from, Poincaré continues, are sterile. The challenge in mathematical creation is to choose the rare few that are "the most fruitful of all" (p. 181).

Poincaré admits that the word "choice" is inadequate in depicting what is at play in this process of mathematical creation. The possibilities to "choose" from are "...so numerous that a whole lifetime would not suffice to examine them" (Poincaré, 1913, p. 181). Yet, those infinites, both in number and sterility, are never consciously considered. The ones that do present themselves to the mind of the inventor are the few possibly fruitful ones that have passed a screening in a "previous examination," and the "inventor were an examiner for the second degree" (Poincaré, 1913, p. 181).

At this point, he invites us to dive deeper into the "very soul of the mathematician" and to realize that "what is important for the psychologist is not the theorem but the circumstances" (Poincaré, 1913, p. 181). As to the fruitfulness of the theorem, mathematicians since have borne sufficient witness. The circumstances, however, have not made a sufficient impact on the understanding of the contemporary psychologist.

In the development of Fuchsian functions, Poincaré describes three instances. In the first instance, he worked for fifteen days "to prove that there could not be any functions like those I have since called Fuchsian functions" (Poincaré, 1913, p. 181). Every combination attempted in this time was fruitless. Then, contrary to his routine, he drank black coffee and could not sleep. "Ideas rose in crowds; I *felt* (emphasis added) them collide until pairs interlocked, so to speak, making stable combinations" (p. 181). The description alludes to the autonomous character of the ideas with Poincaré's role in this particular phase being that of a passive recipient. The next morning all Poincaré had to do was write out the results. Poincaré then describes traveling on a geological excursion leaving his mathematical work behind:

> At the moment when I put my foot on the step (of the omnibus) the idea came to me, without anything in my former thoughts seeming to have paved the way for it, that the transformations I had used to define the Fuchsian functions were identical with those of non-Euclidean geometry.
>
> *Poincaré, 1913, p. 181*

Poincaré did not pursue the insight any further and continued the conversation he was engaged in prior to the intuition. He describes feeling a "perfect certainty" that when he reached back home from the travel, he was able to verify at his leisure (Poincaré, 1913, p. 182).

In the second instance, he describes working on another set of arithmetic questions to no avail. His frustration leads him to take a break from his work and go to the seaside for some time:

> One morning, walking on the bluff, the *idea came to me*, with just the same characteristics of *brevity, suddenness and immediate certainty*, that the arithmetic transformations of indeterminate ternary quadratic forms were identical with those of non-Euclidean geometry.
>
> *Emphasis added, p. 182*

As before, he systematically wrote out his results and was able to do so as effortlessly as before, except he was confronted with another impasse. This third instance of being stuck was also resolved when he took a break from his usual work and engaged elsewhere. This time he was undergoing his military service and walking along the street one day, "the solution of the difficulty which had stopped me suddenly appeared to me" (Poincaré, 1913, p. 182). He did not have to write them out or work on them at that point. After his service was over and when he returned home, he was able to write from memory effortlessly.

The process of mathematical creation can be seen as three distinct phases. The central phase of "sudden inspirations" is sandwiched between two periods of conscious effort. The first phase of conscious effort appears to be entirely in vain. The combinations attempted seem to be "totally astray." Nevertheless, this committed voluntary effort is indispensable for the sudden inspirations. It seems to, as Poincaré (1913) opines, "set agoing the unconscious machine" (p. 183).

The second phase of the conscious effort involves putting "in shape the results of this inspiration, to deduce from them the immediate consequences, to arrange them, to word the demonstrations, but above all is verification necessary" (Poincaré, 1913, p. 183).

Poincaré (1913) uses various terms to describe the central period of "inspiration": "revelation," "intuition," and "illumination" (pp. 181–183). To summarize, these inspirations are experienced as sudden, brief, and autonomous. This sudden appearance into consciousness is a sign that a "long, *unconscious* (emphasis added) prior work" has already taken place (Poincaré, 1913, p. 182). In addition to the first phase of conscious effort, the break in the routine and a conscious engagement in entirely different tasks seem to be typical preconditions. The unconscious work appears to have gone on in this period. The inspirations are usually "felt rather than formulated" and are accompanied by a feeling of "perfect certainty" (p. 182). However, they are not entirely infallible:

> But do not think this a rule without exception; often this feeling deceives us without being any the less vivid, and we only find it out when we seek to put on foot the demonstration. Hence the necessity of the second phase of conscious work to verify the results.
>
> *Poincaré, 1913, p. 183*

Based on these observations, Poincaré (1913) hypothesizes that the unconscious self "is in no way inferior to the conscious self; it is not purely automatic; it is capable of discernment; it has tact, delicacy; it knows how to choose, to divine" (p. 183).

Why, though, Poincaré (1913), asks, of all the infinite (and sterile) combinations do the most useful cross the threshold to conscious appreciation? It is because the most useful is the one that represents the available, disparate facts in a harmony that appeals to the "esthetic sensibility" of the human being (pp. 184–185).

Despite describing this phenomenon as "mathematical creation," Poincaré (1913) notes, "these are generalities applicable in sum to all the sciences; and for example, the mechanism of mathematical invention does not differ sensibly from the mechanism of invention in general" (p. 168).

It is worth noting that the most detailed analysis of Poincaré's description of mathematical creation was carried forward not by a psychologist but by Jacques Hadamard, a mathematician himself. Hadamard (1954) seconds Poincaré's process identifying that his own process also included a wordless stage in the form of mental images (pp. 75–77). He cites examples of Einstein (p. 142), Helmholtz, Gauss, Lamartine, and Mozart (pp. 15–17) as corroborating evidence for what Poincaré has described.

Also noteworthy is Verhulst's (2012) insistence that the phenomenon is that of invention rather than discovery (p. 6). He cites Hadamard as preferring the translation "invention" in the title of his book rather than "discovery," whereas the very first line of Hadamard's personally approved translation states, "we speak of invention: it would be more correct to speak of discovery" (Hadamard, 1945, p. xi). Similarly curious is Toulhouse's comment that "this method of work is not common in science matters and constitutes a very special character of mental activity of M. H Poincaré" (Ginoux & Gérini, 2014, p. 19).

Henri Poincaré—The Philosopher

It is clear from the previous section that Poincaré's analysis of his own process in mathematical creation has philosophical implications in addition to psychological. In the rest of that book, *Science and Method*, the philosophical implications are explored further. Additionally, his publications related to philosophy were published as three compilations: *La science et 'hypothese, La valeur de la science, and Science et methode* in his lifetime with two other publications coming out posthumously, the last one as late as 2002 (Roller/Poincaré, 2002, pp. ix–x). Only the first three were translated into English as *Science and Hypothesis, The Value of Science*, and *Science and Method*. These three are compiled under a common title, *Foundations of Science*, referenced in this paper.

Poincaré's repute as a philosopher of science is received in contrasting ways. On one end of the spectrum, Russell appreciates Poincaré as a mathematician and physicist for his "power of co-ordinating the whole domain of mathematics and physics in a single system of ideas," but then stating, "but these merits, great as they are, are accompanied by what cannot but appear as defects to anyone accustomed to philosophy" (Russel, 1905, p. 412). On the other hand, Royce submits in the preface for *Science and Hypothesis*, "The treatise of a master needs no commendation through the words of a mere learner" (Royce, 1905, p. xv).

Royce (1905) goes on to summarize one of the main contributions of Poincaré's work and illustrates its significance with examples from other fields. He acknowledges that as science advances in its control over nature and "in its details more assured" (p. xvi), "there is no longer any notable scientific orthodoxy" (pp. xvi–xvii). The certitude that could be attributed to the most "general theories," the ones held to be most fundamental in science, can no longer be held as rigidly as may have been earlier warranted.

Does this imply, Royce (1905) goes on to inquire, that we do away with theorizing at the most general level altogether (p. xix)? In postmodern terms, are worldviews and metanarratives entirely useless and untenable? Royce answers in the negative and sees Poincaré's work as instrumental in providing a more nuanced response to the dilemma:

> But in the main, this volume must be regarded as what its title indicates—a critique of the nature and place of hypothesis in the work of science and a study of the logical relations of theory and fact. The result of the book is a substantial justification of the scientific utility of theoretical construction—an abandonment of dogma, but a vindication of the rights of the constructive reason.
>
> *p. xxi*

Indeed, we see that Poincaré (1913) avoids the naive certainty of early modernism and the reactive relativism of postmodernism (what Griffin identifies as "deconstructive postmodernism" (Griffin, 2000, p. x). Poincaré states:

> On a little more reflection it was perceived how great a place hypothesis occupies; that the mathematician cannot do without it, still less the experimenter. And then it was doubted if all these constructions were really solid and believed that a breath would overthrow them. To be skeptical in this fashion is still to be superficial. *To doubt everything and to believe everything are two equally convenient solutions; each saves us from thinking* (emphasis added).
>
> *Poincaré, 1913, p. 12*

Royce (1905) sees Poincaré's nuanced response to the highlighted issue in how he classifies hypotheses in two broad categories:

1. The hypotheses which are valuable precisely because they are either verifiable or else refutable through a definite appeal to the tests furnished by experience; and
2. The hypotheses which, despite the fact that experience suggests them, are valuable despite, or even because, of the fact that experience can neither confirm nor refute them. The contrast between these two kinds of hypotheses is a prominent topic of our author's discussion.

p. xxi

The first category of hypotheses is well known, and their utility is understood in the scientific enterprise. However, the second class of hypotheses is not given its due regard. In understanding their significance and relationship to the first category of hypotheses, we would be able to appreciate Poincaré's contribution. The second class of hypotheses, suggested by experience but can't be confirmed or refuted thereby, places science as an interpretation of facts of nature rather than a "portrayal or a prediction" thereof (Royce, 1905, p. xxii). The dogmatic approach, more characteristic of early modernism, that looks at science as an unambiguous representation of reality can only sustain itself by ignoring the presence of this second class of hypotheses. However, given the interpretive nature of this class of hypotheses and, therefore its fallibility, the question remains to its utility and its possible arbitrariness.

Royce (1905) elaborates how important it is that "experience suggests" the formulation of these hypotheses. He contrasts Poincaré's stance with that of Kant. Kant would accord the a priori "ideas of reason" precedence such that they would guide the empirical investigations, but in no way can those ideas be changed considering those investigations. Instead, the results of the investigations have to be seen in light of the rigidly predetermining ideas. On the other hand, for Poincaré:

> ...all this adjustment of our interpretations of experience to the needs of our intellect is something far less rigid and unalterable, and is constantly subject to the suggestions of experience. We must indeed interpret in our own way; but our way is itself only relatively determinate; it is essentially more or less plastic; other interpretations of experience are conceivable.
>
> *p. xxiii*

These hypotheses are the most convenient aids to understanding experience, but "this convenience is not absolute necessity" (Royce, 1905, p. xxiii). Experience has a vital role in influencing ideas that will prove most convenient.

We see then that neither is science a mirror reflecting experience as it is nor is it a projection of the a priori conclusions of the scientist. As Royce (1905) states:

> Instead of Kant's rigid list of a priori forms,' we consequently have in M. Poincaré's account a set of conventions, neither wholly subjective and arbitrary, nor yet imposed upon us unambiguously by the external compulsion of experience.
>
> *p. xiii*

The question remains as to why these hypotheses are necessary. Royce (1905) shows that the pragmatic value of this second class of hypotheses is in how they serve as "leading ideas" (p. xxix) and in how they have "proved extremely fruitful in guiding research" (p. xxvii). Without these leading ideas, developing the refutable or confirmable first class of hypotheses is not possible. And without these, empirical science is not possible. Poincaré expounds the specific details in the text concerning the principles of mechanics and theory of energy (Royce, 1905,

p. xxvii). Royce (1905) himself illustrates the validity of this conception when seen in fields as disparate as history and microbiology.

This understanding of hypotheses leads us to understand the dialogical nature of science. The scientist, as interpreter, is letting "the facts speak" but is also in a position to "talk back" to the facts:

> Man is not merely made for science, but science is made for man. It expresses his deepest intellectual needs, as well as his careful observations. It is an effort to bring internal meanings into harmony with external verifications. It attempts therefore to control, as well as to submit, to conceive with rational unity, as well as to accept data. Its arts are those directed towards self-possession as well as towards an imitation of the outer reality which we find. It seeks therefore a disciplined freedom of thought.
>
> *Royce, 1905, p. xxvi*

Additionally, we can see a form of pluralism that the possibility of "other interpretations of experience" opens the doors to. A dogmatic certitude allowed for only one relationship with alternate interpretations: blanket rejection. Similarly, ignoring hypotheses at this level altogether makes for a shallow eclecticism. Poincaré's work paves the way for a meaningful dialogue between scientists and facts of nature *and* between scientists holding different interpretations. The "community of inquirers" can thus develop a more nuanced relationship. One that allows the different partners in dialogue to offer a mutual corrective and simultaneously grow toward a better understanding and contribution toward the evergrowing facts of reality and their relationship.

Conclusions, Limitations, and Further Lines of Inquiry

This paper sought to show a few dimensions of the *unitas multiplex* (Stern, 1938, p. 73) that Henri Poincaré was. It looked at how different aspects of his personal life impacted his pursuit of science and success therein. His reflection over his own work as a scientist led him to offer critical insights for psychology of science. Further, his undogmatic approach to science allowed him to contribute to promising lines of inquiry within the philosophy of science.

As unmusical as I am in mathematics and physics, this paper reflected that lack of understanding. In Poincaré's analysis, I neither have the intuitive capacity nor the memory for mathematical understanding. A far more holistic approach to Poincaré's life and work would have to include at least a basic understanding of his scientific contributions, particularly concerning the trends in the fields in his age. Russell's critique of Poincaré, in contrast with Royce's appreciation, may also then be better understood.

It will be helpful to look at Poincaré's work from the perspective of Royce and those of the other pragmatists. For example, the Roycean triad of beauty, duty, and truth could offer a valuable framework to tie in Poincaré's work in the *Value of Science* (corresponding to duty), along with what was seen of his work in this paper in *Science and Hypothesis* (corresponding to truth) and *Science and Method* (corresponding to beauty). Further, the implications of Royce's work for epistemology (for example, looked at in relation with John Dewey's *Reconstruction in Philosophy* and Peirce's *Notes on Scientific Philosophy*), for ontology (for example, looked at in connection with Viktor Frankl's Dimensional Ontology), theology (Peirce's treatment of abduction in *A Neglected Argument for the Reality of God*), and psychopathology (in light of the manifestation of the intuitive in John Nash's science as compared to his possible schizophrenia).

QUESTIONS FOR DISCUSSION

1. Intuition is sometimes contrasted with the formal reasoning central to the scientific method. How does Poincaré challenge this, and how might we think of the role of intuition in relation to scientific reasoning?
2. Why is the distinction between different forms of hypothesis important to science? Does psychology take this into account in its own methods?
3. How does Poincaré's description of his discovery process implicate both philosophy and psychology?

References

Eysenck, H. J. (1995). *Genius: Natural history of creativity*. Cambridge: Cambridge University Press.

Ginoux, J., & Gérini, C. (2014). *Henri Poincaré: A biography through the daily papers*. New Jersey: World Scientific.

Griffin, D. R. (2000). *Religion and scientific naturalism: Overcoming the conflicts*. Albany: State University of New York Press.

Hadamard, J. (1954). *The psychology of invention in the mathematical field*. New York: Dover.

Poincaré, H. (1913). *The foundations of science*. New York: The Science Press.

Poincaré, H. (2002). *Scientific opportunism an anthology L' opportunisme scientifique* (L. Rollet, Ed.). Basel: Birkhäuser.

Royce, J. (1913). Preface. In H. Poincaré (Ed.), *The foundations of science* (pp. 4–12). New York: The Science Press.

Russell, B. (1905). Review of science and hypothesis by H. Poincaré. *Mind, 14*, 412–418.

Stern, W. (1938). *General psychology from the personalistic standpoint* (H.D. Spoerl, Trans.). New York: Macmillan. (Original work published 1935.)

Verhulst, F. (2012). *Henri Poincaré: Impatient genius*. New York: Springer.

Weinstein, G. (2012). TY – JOUR AU – Weinstein, Galina PY – 2012/07/03 SP – T1 – A Biography of Henri Poincaré – 2012 Centenary of the Death of Poincaré ER.

5

WHEN THE MIND CANNOT BE TRUSTED

The Lonely Genius of John Nash Jr.

Ron C. Hopkins

Personal Preamble

Like many, I was first introduced to John Nash Jr. through Sylvia Nasar's *A Beautiful Mind* (Nasar, 1998). Before reading her masterfully written biography, my interest in mathematics was minimal at best. I was far more interested in the psychology of beliefs—how they are formed and how they influence the way we understand ourselves and the world around us. Nash's story caught my attention because of an answer he gave in an interview. Nash had suffered from auditory hallucinations caused by his schizophrenia, and he came to believe he was receiving messages from extraterrestrials. Later in life, when his symptoms were largely in remission, he was asked why a mathematician devoted to reason and logic could believe that extraterrestrials were sending him messages. His response? "Because the ideas I had about supernatural beings came to me the same way my mathematical ideas did" (Nasar, 1998, p. 11). This started my fascination with Nash's work and life.

The scientific genius often functions as a trailblazer, forging paths that only they can see. Theories develop on the foundation of collected evidence, but they first began as a belief. A singular conviction is held before evidence can be collected. It is this time between the initial idea and accepted scientific law where the genius can be most alone. Here was a person trying to make sense of the world only to be betrayed by his sense-making. It is this aspect of Nash's life that motivates my own work in studying depressive cognition.

One of the most overlooked aspects of depression is how isolating it can be for the individual. A depressive state can turn you against yourself. You suddenly find yourself without the energy to rise from bed. Joy and pleasure are leeched out of your life and you doubt they will ever return. You have no idea how or why this is happening to you. You may try to reach out for help only to discover

DOI: 10.4324/9781003276692-7

an insurmountable gulf of understanding between yourself and others. The most unnerving and debilitating feature of depression is how it strands you in a place that only you can inhabit. While depression and schizophrenia are two very different mental disorders, they share the same tragic feature of alienation and imprisonment in the mind. It is this aspect of Nash's struggles that I do not think receives as much attention as it should. His experiences show us the complexity of our individual mental lives. Nash's work changed the way we understand human behavior. I hope his life story would also encourage us to understand the lived experience of mental disorders beyond mere symptoms and diagnostic criteria.

Introduction

This chapter presents a case study of the late American mathematician John Nash, Jr. (1928–2015). Sylvia Nasar's text remains the most thorough and detailed account of Nash's life currently available. For this reason, it was the primary source of biographical information. However, wherever possible, primary sources were used to have Nash's own words and thoughts inform the study. This included previously recorded interviews, speeches, and autobiographical sections published after Nasar's 1998 biography.

The case study, informed by the theories and research aims of psycho-biographical methods, is intended to consider the nature of the thought patterns experienced by John Nash, Jr. within the context of his personal, professional, and social development, with an emphasis on how they relate to his work as a mathematician. Because of its compatibility with Merleau-Ponty's (1964) onto-logical philosophy, which emphasized the interconnectedness between the indi-vidual in the world and the world in the individual, Levinson's (1986, 1996) theory of the human life cycle was used to uncover and reconstruct Nash's development throughout his lifespan. The specific approach employed in this case study is taken in part from a similar psychobiography of the life of Steve Jobs by Fouché et al. (2017). In the case of Nash, this interconnectedness remains woefully underexplored, possibly due to the perception of Nash's irrationality due to his mental illness. That is, the stigma of mental illness surrounding Nash's story may contribute to a devaluating and dismissal of the particular style of rationality he employed. The author hopes that this current study presents a con-vincing case for how Nash's complex life and experience could inspire future studies instead of curtailing such investigation.

We should note first that Nash's life has been the subject of considerable attention and discussion, with the focus evenly split between his contributions to game theory (see Narahari, 2016 for a summary) and his struggles with mental health (Nasar, 1998; Andreasen & Nash, 2015). The former gained considerably more attention after the 1998 publication of *A Beautiful Mind*, a biography of his life that gained critical acclaim (Nasar, 1998). Three years later, a film adaptation of the book would win multiple Oscars, including Best Picture. The film was accused of changing many details regarding Nash's life. Some of these were more justified, such as showing Nash benefiting from medication for his condition (see Suellentrop, 2001; von Tunzelmann, 2012). Nash resisted taking medication in real life as he believed it impaired his thinking. Other film changes were less justified, such as changing the nature of his hallucinations and ending the film with Nash delivering a speech at the Nobel ceremony in which he attributes his recovery to the power of love (Howard, 2001, 1:45:13). The real Nash never delivered a speech at the ceremony, let alone voiced any such opinion. Yet the fictitious quote is still attributed to him in popular culture to this day ("19 John Nash Quotes That Will Inspire You," 2021). The resulting publicity from both book and film resulted in competing narratives concerning Nash's life: The Hollywood version gave us a version of Nash whose life fit perfectly within the standard three-act narrative, complete with a happy ending that affirmed the power of love to cure all ills. The Nash version came directly from the man himself, in which he claimed to be cured of schizophrenia later in life, despite psychiatrists who met him in person noting he still demonstrated symptoms (Andreasen & Nash, 2015). To these versions we may add the myriad of other narratives that have tried to make sense of Nash's life's overlapping and sometimes contradictory accounts. There remains an important psychological need to better understand the nature of a mind that can produce both genius and chaos as both remain woefully underdefined and misunderstood. This is an aspect of the psychology of science that also stands in need of more attention.

Mental Disorders and the Ethics of Psychobiography

The relationship between "genius" and "madness" has been the subject of considerable interest to psychology and other fields. This has ranged from debunking the "mythconception" of the mad genius (Dietrich, 2014, p. 1) to the neuroscientific exploration of differences between normal and abnormal functioning in the brain (see Buckner, 2013 for a summary of prevailing theories). One approach with potential to yield important insights is the method of psychobiography, which focuses on the life and experience of persons with special abilities. For example, this approach has been used with regard to chess master Bobby Fischer (Ponterotto & Reynolds, 2013), musician Brian Wilson (Belli, 2009), Helen Keller (van Genechten, 2009), and Steve Jobs (Fouche et al., 2017). However, using this method raises several ethical concerns which must be acknowledged. The impact of psychologists' evaluative judgments may considerably affect both other professionals and the public (Ponterotto & Reynolds, 2017), especially when psychopathology is involved. A 2017 study compared the presentation of psychopathological information with psychobiographical information and found a significant impact in changing attitudes of individuals toward mentally ill people (Kang & Son, 2017). The authors argue that negative attitudes held toward individuals with mental disorders often stem not from the disorder but from social perceptions of the disorder.

"The negative image is given [to the person with the mental illness] by society rather than the trait itself" (Kang & Son, 2017, p. 452). The study observed significant change when participants were provided with psychobiographical information rather than merely psychopathological information. In other words, attitudes about mental disorders changed once participants were learning about the individual as opposed to only learning about their symptoms. This suggests that the methods of psychobiography may provide substantial benefits to the clinical fields, supporting efforts to reduce stigma and increase the effectiveness of patient–client relationships. Negative social perceptions of individuals with mental disorders can impede the formation of supportive social networks as well as discourage the individual from seeking out professional help. Put simply, I would argue that the goal of psychobiography in the context of mental health is ultimately to reduce the "otherness" of those suffering from mental disorders.

Nevertheless, it is important to describe the nature of the disorder in question as it is currently understood within psychology, and how it might relate to creative thought. Previous studies have suggested a connection between creative potential and the presence of affective disorders (Hare, 1987; Andreasen, 1987; Ludwig, 1995) and the potential relationship between creativity and various psychotic traits (Brod, 1997). A 2010 study found that the dopamine system in creative individuals shared many similarities with those found in people with schizophrenia (de Manzano et al., 2010). A correlation between mathematical ability and mental illness has been the subject of some debate. A possible relationship between psychosis and high performance has been observed in studies (see Karlsson, 2004), and the history of the mathematics field has no shortage of examples. Pythagoras, Isaac Newton, Kurt Godel, Ludwig Boltzmann, Alexander Grothendieck, and Florence Nightingale all succumbed to mental illnesses later in their lives (see Burton, 2010).

Additionally, Nash was not the only genius mathematician who suffered from schizophrenia. Vashishtha Narayan Singh's own struggles with the illness shared many parallels with Nash, with the tragic exception of recovery later in life (Burton, 2010). Potential causes or explanations for these cases are varied, and to date, there is very little compelling empirical evidence to support any specific theory. However, of interest to this current study is the growing neuroscience evidence that suggests that creativity emerges from states in which unusual patterns of connectivity are observed between the executive control network and the default mode network the executive control network is viewed as largely responsible for cognitive inhibition (Beaty et al., 2014).

In contrast, the default mode network is believed to contribute to internal modes of cognition used in recall, thinking about the future, and mind wandering (Buckner, 2013). According to this model, mental health is primarily a result of the two networks operating in dynamic tension: "Genius comes from blending the two sides. The problem is managing the blend while maintaining stability" (Jack, n.d., as cited in Perina, 2017). Evidence from neuroimaging studies does suggest that schizophrenia may be the result of aberrant brain connectivity at the network level (Friston, 1998; Raichle et al., 2001; Garrity et al., 2007; Whitfield-Gabrieli et al., 2009). Consequently, "constant over-engagement of the default network could lead to an exaggerated focus on one's thoughts and feelings and an ambiguous integration between one's thoughts and feelings with events in the environment" (Whitfield-Gabrieli et al., 2009, p. 1283).

This model can be useful for framing discussion of Nash's mental health, and help to counter the tendency to fall into the trap of a binary representation of sane/insane, healthy/sick, or rational/irrational. Nevertheless, cognitive models do not provide a sense of how Nash, the person, himself experienced his world and even his thought processes. Therefore, for the purposes of this chapter, Nash's mental state(s) are observed as reported, as the research aims of this study are to take up Nash's life as he experienced it.

"Mathematical Madness" and Genius

Any discussion of Nash's mental struggles in relation to his mathematical brilliance should note that Nash was not the first mathematician to struggle with mental health. Pythagoras, Isaac Newton, Kurt Godel, Ludwig Blotzmann, Florence Nightingale, Vashishtha Narayan Singh, and Alexander Grothendieck are notable examples of brilliant minds who later struggled with different types of psychopathology. Much like Nash, C.P. Ramanujam also suffered from schizophrenia. Tragically, Ramanujam took his own life at the age of thirty-six.

The idea that genius and madness are often intertwined is not merely a romantic notion. Psychotic spectrum disorders are disproportionately diagnosed in highly creative individuals (Kaufman & Paul, 2014; Greenwood, 2016). Creativity and creative genius are an underexamined aspect of scientific progress, despite their often being the catalyst for advancement in many fields (see Gruber & Bödeker, 2005). As the philosopher Kant posited, "Genius is a talent…for producing that for which no definite rule can be given" (Kant & Meredith, 2007, p. 46). In other words, a genius is one who can move beyond the current theories and models. Thus, we might say that genius sees what no one else is capable of seeing…*yet*. It is this aspect of the genius which I would argue is critical for understanding Nash's mind. We may even call this the "epistemic loneliness" of the trailblazing intellect—persistently beyond the rest of us, always waiting for us to catch up to them. Consequently, I would argue that this perspective should frame any exploration of Nash's life.

Descriptive Understanding of John Nash, Jr.

Most of the biographical information presented here comes from Sylvia Nasar's 1998 biography of John Nash. However, the following biographical account has been structured and organized for comparison to the Levinson model.

Pre-Adulthood Stage (Age 0–22)

In the Levinson theory, pre-adulthood comprises four stages: early childhood, middle childhood, adolescence, and the early adult transition (Levinson et al., 1978). This era is characterized by rapid biopsychological growth and individuation (changes in a person's relationship to themselves and the external world) from infant dependence toward progressively more adult independence (Dahlberg, 2006). The process of individuation is one in which the person begins to distinguish themselves from the "not-me(s)" surrounding them (Levinson et al., 1978). The individual's social world expands from the immediate family to the larger sphere of school and peer groups around the ages of five or six. The transition from middle childhood to adolescence usually occurs around ages twelve and thirteen. According to Levinson's theory, adolescence is the conclusion of the era of pre-adulthood.

Nash's Pre-Adulthood Stage (1928–1950)

John Nash, Jr. was born on June 13, 1928, in Bluefield, West Virginia, to parents John and Virginia Nash. He had one sibling, Martha, born two years later in 1930. The elder Nash was an electrical engineer for the Appalachian Electric Power Company, while Virginia had been a schoolteacher before marriage. The Nashes were financially secure, as John, Sr.'s steady income was enough to keep the entire family in the upper-middle class (Nasar, 1998). Even during the Great Depression, the Nashes were still attending high society social gatherings. Both parents took considerable interest in their children, seeing them as their highest priority. John (Sr.) would take his kids to his company's power lines for weekend inspections. He would use these opportunities to teach his children firsthand about topics such as electricity, geology, weather, and astronomy. He did not believe in talking down to his children and instead always interacted with them as if they were adults. When John (Jr.) began outperforming his classmates in school, his mother began teaching him at home rather than stunt his learning by forcing him to stay at the same level as other children of his age.

John (Jr.) was a solitary child who preferred his own company. He was quite intelligent, being able to read by the age of four and becoming a voracious reader by the age of seven. Socially, he was awkward and prone to acting strangely around other children of his age. This was a stark contrast to Martha, who was very sociable. She was known to create elaborate games and play with other kids. She would invite John (Jr.) to join them, but he resisted. He was awkward around other kids his age, to the point where his quite sociable sister had difficulty engaging with him. This difficulty eventually led to a noticeable lack of affection between the siblings. By the age of twelve, he had put together his own makeshift laboratory in his bedroom to conduct his scientific experiments.

Nash was enrolled at Bluefield High School, rejoining formal education after spending much of his early social development years isolated from others. The radical shift in both learning and socializing was quite difficult. He performed poorly in the classroom, receiving poor grades despite being quite intelligent. The worse he performed in school, the more he sought knowledge elsewhere. He was known to read the encyclopedia before going to bed at night.

Socially, Nash was known for having a "big mouth," which he used to monopolize classroom conversation, even during lectures. He was a social outcast who responded to ostracization with practical jokes and demonstrations of his intellectual superiority. For example, while still in high school, Nash attended classes at Bluefield College. He famously attended a magic show at a Bluefield Carnival and was unimpressed with the magician's trick of creating sparks between two swords. When challenged by the magician to perform the same trick, Nash did so easily. "It's nothing to it…it's just static electricity," he remarked to the awed crowd (Nasar, 1998). His classmates responded to his retaliatory behaviors by escalating to bullying him. He was known to have a severe temper, and his peers would often goad him into fights, knowing he would lose and be further humiliated.

Nash would eventually develop a close-knit group of friends, drawn together with a shared love of experimentation with homemade explosives. However, these experiments ended when Herman Kirchner accidentally detonated a pipe bomb and died *en route* to the hospital. Nash's parents were largely unaware of these events, even after the death of Kirchner. Nash's love for pranks did not diminish and continued well into adulthood.

Upon graduation from high school, Nash attended the Carnegie Institute of Technology (now Carnegie Mellon University) on a Westinghouse scholarship. He first pursued chemical engineering and chemistry but struggled with technical drawing and laboratory work. He switched to mathematics and instantly flourished. Nash would later credit E.T. Bell's book, *Men of Mathematics*, as the inspiration for his interest in mathematics. Notably, this book focused on the anecdotes and achievements of the men profiled and was not a rigorous history of the field (Bell, 2008/1937). Nash's interest in becoming a mathematician would often be attributed to his strong desire for fame and recognition (Nasar, 1998). This desire was a vital component in his motivation for professional success. Nash advanced so quickly through the mathematics program that he was awarded his Master's degree and Bachelor's degree simultaneously.

Socially, Nash was respected for his intellect but avoided by his classmates due to his odd behaviors. His humor was off-putting, awkward, and made people uneasy. It was generally accepted that Nash was a genius with a bright future, and therefore many of his behaviors were excused as the eccentricities of a brilliant mind. Nevertheless, he had great difficulty engaging with his peers, often appearing to savor his "outsider" position both academically and socially (Nasar, 1998, pp. 25–32; Kuhn & Nasar, 2007, pp. 5–6).

Early Adult Transition (Age 17–22)

Levinson's theory holds that the transition to early adulthood occurs from 17 to 22. Three main developmental tasks happen during each transitional person: the ending of the existing life structure, individuation, and the initiation of a new structure (Levinson et al., 1978; Dahlberg, 2006). This is when the individual creates an initial base for adult life.

Nash's Early Adult Transition (1945–1951)

For Nash, this was a period of accelerated professional development. He was recognized for his genius and had a pronounced desire for fame and notoriety. In his engagements with other students, he was callous toward any person he deemed less intelligent than himself (Nasar, 1998). If any "less intelligent" persons disagreed with him, he would react angrily. Nevertheless, Nash succeeded in completing the Levinsonian tasks of the early adult transition.

Early Adulthood Stage (Age 22–28)

The individual has two tasks during the period of entry into adulthood (Levinson, 1978). The first is to explore the possibilities of adult life by maximizing alternatives and avoiding strong commitments. The second is to create a stable life structure and become more responsible. Levinson proposes that this is the period during which individuals must find a balance between exploring the adult world while building a stable life (Levinson, 1978, pp. 57–58).

Nash's Early Adulthood Stage (1951–1957)

When Nash applied to Princeton's mathematics department, his former advisor R.J. Duffin wrote a single sentence recommendation for him. It stated, "This man is a genius" (Nasar, 1998). Nash was ultimately accepted into both Harvard and Princeton. After being aggressively pursued by the chairman of the mathematics department, he chose Princeton, which offered him a prestigious John S. Kennedy fellowship. The Princeton offer convinced him that the school valued him more than Harvard (Nasar, 1998; Nash-Jr., 1994).

While at Princeton, he became obsessed with games and game theory, a topic of great interest in math departments during the Cold War. He created his own games, such as the "Game of Nash" (Nasar, 1998), which became quite popular on campus and even drew the attention of the legendary mathematician John von Neumann. One game created by Nash would later be published as So Long Sucker. The game was designed, so players are forced to make and then break agreements with other players, hence the name. Although before publication, Nash's original name for the game was F**k you, Buddy! (Hausner et al., 1964).

Like his pranks, Nash's engagement with other peers was primarily to provoke strong responses. The parallels between such behavior and his eventual area of mathematical expertise, game theory, are difficult to ignore.

Game theory is essentially an analysis of decision-making. It focuses on how people make decisions from a series of finite choices to arrive at the best possible outcome for themselves. Thus, game theory is primarily a theory of human behaviors. As it would come to be called, the Nash equilibrium was Nash's significant contribution to the field. His theory shifted game theory in economics away from the dominant two-person zero-sum theory, which held that one person's gain is the other person's loss. Nash's equilibrium focused on a "configuration of strategies, such that no player acting on his own can change his strategy to achieve a better outcome for himself" (Nasar, 1998). This approach would eventually be applied to nuclear conflicts between world superpowers. There is a notable irony in the antisocial Nash arriving at an explanation for human interaction, which was far more accurate than the prevailing theories. Part of this may lie in Nash's unique mental method of approaching mathematics.

While outside of Levinson's developmental stages theory, it is worth noting that Nash was an intuitionist in his mathematical thinking. His genius was not due to his mind working faster or retaining more information. His insights came from nonrational flashes of intuition more comparable to the arts than is most often associated with the gradual inferences of scientific reasoning (Nasar, 1998). Other mathematical intuitionists functioned in the same manner. Just like Nash, Georg Friedrich Bernhard Riemann, Jules Henri Poincaré, and Srinivasa Ramanujan would have the vision first and then set about creating the proofs afterward. This approach to mathematics would entail both right and left brain hemispheres working with considerable communication between them (Capps, 2011). This point has significant relevance for the later stages of his life, to be discussed in the following sections.

During this time, Nash's development was arguably an inverse of the Levinsonian model. He invested all his time and efforts into the singular pursuit of professional success. For him, adulthood was about the pursuit of fulfilling his ambitions at the expense of creating a more stable life structure for himself. This is observed in his first real romantic relationship, which began in 1952 when he was 27. By this time, he had graduated with his Ph.D. and worked as an instructor at the Massachusetts Institute of Technology (MIT). While getting treated for varicose veins, he met nurse Eleanor Stier. The two would go on to have a child named David. Preferring to spend his time among those he considered intellectually worthy, Nash ignored Eleanor and their child. He told Eleanor that an MIT professor should marry someone of equal intelligence and not a nurse with inferior intelligence (Nasar, 1998). Eleanor and David lived in squalor as Nash refused to support them financially. Eleanor contacted Nash (Sr.), and when he learned of the situation, he called Nash (Jr.) and berated him for his treatment of the mother of his child. Nash (Sr.)

told him to marry Eleanor, but Nash (Jr.) refused, opting instead to try and adopt his son. Eleanor rebuffed the offer. Soon afterward, he met Alicia Larde Lopez-Harrison, a physics graduate from MIT. The two married in 1957.

The Age 30 Transition (Age 28–33)

According to Levinson, this transitional period is when the individual is provided with an opportunity to work on the flaws of the initial adult life structures to create a more satisfactory one (Levinson et al., 1978). This transition may be relatively smooth or result in a developmental crisis if individuals find their current life structure unsatisfactory (Levinson, 1996).

John Nash's Age 30 Transition (1957–1962)

Nash began first exhibiting symptoms of mental problems around the age of 30. Despite having no family history of mental disorders and demonstrating no signs of declining mental health up until this point, he began behaving erratically and growing paranoid (Nasar, 1998). Nash himself said the issues started around the time Alicia became pregnant with their first child. He began demonstrating increasingly bizarre behavior, such as proclaiming that his photograph on the cover of Life magazine had been disguised to look like Pope John XXIII, complaining that aliens were ruining his career from outer space, or that his room at home was bugged by the government (Nasar, 1998). He delivered a rambling, barely coherent lecture at the American Mathematical Society deemed. Mathematician and friend Donald Newman recounted: "Everybody knew something was wrong. He didn't get stuck. It was his chatter. The math was just lunacy" (Nasar, 1998, p. 247). Yet, much like in his undergraduate and graduate years, these odd behaviors were tolerated or at least ignored by those around him, attributing them to eccentricities of genius.

Despite his odd and paranoid behavior, he continued to work regularly until 1959. While giving a lecture on the Riemann hypothesis at Columbia University, he became incomprehensible in his speech. Realizing he needed professional help, he was taken to the McLean Hospital part of Harvard Medical School. However, he was very reluctant to accept treatment as he refused to acknowledge that he was even experiencing delusions and he did not like how antipsychotic medications made him feel lethargic and slow. It was the first of many episodes of hospitalization during the 1960s. In a later interview, Nash would pointedly explain:

> I never went voluntarily…I didn't feel that I belonged locked up… In madness, I thought I had a very important role, and, of course, that includes the messenger-type function. That is a Muslim concept particularly with

Muhammad. He's the messenger of Allah. That's, I think, a standard phrase. So I saw myself as being a messenger or having a special function. I saw it in terms of there being supporters, but also of opponents, and so I would think that if I was put in the hospital, it would be a coup d'etat by the opponents.

Public Broadcasting Service, 2002

Later, when recalling these periods of mental instability, Nash would say he experienced "a feeling of mental exhaustion and depletion, recurring and increasingly pervasive images, and a growing sense of revelation regarding a secret world that others around him were not privy to" (Nasar 1998, p. 244). After 50 days of involuntary commitment, he was released. He immediately took Alicia on an extended European trip, during which he went to Switzerland and attempted to renounce his U.S. citizenship. This act required intervention by the U.S. State Department (Nasar, 1998, pp. 264–265).

When the couple returned in April 1959, they settled in New Jersey so Nash could work at Princeton. Alicia thought that being around other mathematicians again would help Nash recover. It did not. His behavior continued to be erratic, although notably when in the company of graduate students, he would often make self-deprecating jokes about how he knew he was "crazy" (Nasar, 1998). When his condition worsened, his family arranged for him to be committed to Trenton Psychiatric Hospital in January 1961. He was given insulin treatments and shock therapy, the standard course of treatments during this time. This would become the start of decades of Nash going in and out of treatments, always returning to his work whenever his illness appeared to go into remission.

Returning to Nash's intuitionist style of mathematical thinking, it is worth noting that months before his psychotic breakdown, he was working to solve the classic Reimann's Hypothesis, and in fact, Nash's breakdown occurred during a lecture on Reimann's Hypothesis. His intuitionist approach was recognized as a liability more than an advantage in game theory. To quote a mathematician familiar with both the problem and Nash's style of thinking:

> For a person who is not a library hound, it's a very dangerous area to go into. If you have a flash of an idea with a scenario and think you may get a result, in the first flash of illumination, you think you have a revelation. But that's very dangerous.

Nasar, 1998

Nasar suggested that Nash's ambitious compulsion to scale the "most dangerous peak" proved central to his undoing (Nasar, 1998). Nash himself attributed his breakdown to his trying to achieve too much too soon, an interesting point to consider with Levinson's developmental stage theory.

The Culminating Life Structure for Early Adulthood—Settling Down (Age 33–40)

During this period, the individual has two major tasks to perform. The first is to establish a niche in society by developing competence in a chosen craft. The second is to work at building an improved lifestyle and to find affirmation from others (Levinson et al., 1978). The individual eventually becomes a full-fledged adult and a senior member of their own world (Levinson, 1996).

John Nash Jr.'s Culminating Life Structure for Early Adulthood (1962–1969)

During this development period, Nash was once again in nearly the inverse of Levinson's proposed stages. Instead of developing competence in his chosen craft, he had long since established himself as not only being competent but as a trail-blazer. He is regarded as an individual who is *ahead* of most others in the craft, making him an outlier for most developmental models. Tragically, at this Stage, Nash experienced a failure to find affirmation from others. In this way, he is also an outlier as he was now the subject of attention from those trying to "fix" what is wrong with him—what makes him different from everyone else.

In December 1962, Alicia initiated divorce proceedings because Nash resented her and refused to engage in any physical relations with her. They were divorced in August 1963. However, Alicia would continue to arrange care and treatment for him because she still felt responsible for him. She helped arrange treatment with a private mental hospital near Princeton, the Carrier Clinic. He responded well to the treatment, where his psychiatrist believed the recovery was permanent. Nash began making plans to teach a few courses at Princeton. However, in February 1964, Nash began to have bouts of insomnia during which his mind would become "filled with the thought of performing imaginary computations of a meaningless sort" (Nasar, 1998, p. 311). He traveled to Europe on a whim, and while in Rome, standing in front of the Forum, he began to hear voices. In a manic state, he came to believe the voices were those of mathematicians who opposed his ideas and whose words were fed into a central machine that inserted them directly into his brain.

He attempted teaching at different universities over the next several years, taking advantage of times when the worst of his condition went into remission. By 1967, plans for him to teach a course at MIT were canceled due to the overall deterioration of his mental state. During this time, he stopped taking medications because they would silence the voices he heard, and this interfered with his belief that he was now a secret messianic figure who received essential messages through the voices.

From June 1967 to 1969, he lived in his mother's apartment until she passed away. He had become highly delusional, communicating with his former colleagues with bizarre letters and convoluted phone calls. Notably, his delusions shifted from the more grandiose themes of earlier years to a more persecutory tone (Nasar, 1998). In a letter to Eleanor, he wrote that he must "sympathize more with the true needs of liberation...liberation from isolation. I'm a refugee, in fact, from false symbols and dangerous symbols" (Nasar, 1998, p. 328). Following his mother's death, his remaining family members tried to have him committed to the state sanitarium. Still, they were refused because his paranoid ideas did not interfere with his ability to maintain himself. Nash viewed this as an act of betrayal by his sister and severed all connection with her afterward.

During this time, Nash had become obsessed with codes and numerology. Much has been made of this obsession, which preceded his eventual progress toward remission and recovery. Nasar arguably romanticized it as being a case of Nash "looking to the order of numbers when the world falls apart" (Nasar, 1998, p. 333), and his "romance with numerology blossomed when his [internal] world was falling apart" (p. 333). However, an alternate theory provides an interesting counterargument. Capps (2011) argues that poor communication between hemispheres identifies the schizophrenic brain; his interest in numerology demonstrated a slowly rebuilding communication between both hemispheres. To be clear, Capps does not make a casual argument on this point. Instead, he argues that the romance with numerology reflected a shift toward improved regulation (Capps, 2011).

The Midlife Transition (Age 40–45)

According to Levinson, this period of transition is when the individual moves from early to middle adulthood. It is a period of moderate to severe crisis, depending on the individual's circumstances. The neglected parts of the self tend to seek expression, which stimulates modification of the existing life structure (Levinson et al., 1978). The most important task during this period is to take a new step in the process of individuation (Levinson, 1986).

Nash's Midlife Transition (1969–1974)

The administration of Princeton allowed Nash to use the school's facilities as he worked to rebuild his life and career. He would stalk the halls of the campus, writing odd math equations and messages on classroom blackboards. He became known as the "Phantom" on campus, and many stories and myths developed as a result. In 1970, Alicia returned and reconciled with him so he could spend time with her and their son. During this time, he became a person of "relatively moderate behavior" (Nasar, 1998, p. 336) and began to emerge from "irrational thinking" (Nasar, 1998, p. 336). He claimed that this occurred due to the natural hormonal changes of aging instead of medication (which he claimed to have stopped taking). While some symptoms of schizophrenia do appear to reduce later in life, this is not always the case (Folsom et al., 2006).

Concerning Levinson's theories, the outlier aspect of Nash's development again casts him as performing almost an exact inverse of the tasks required by the theorized Stage. Arguably, the individual proceeding through the stages in the manner described has achieved stability which is then threatened by the crisis. It is a period during which the individual's whole is addressed to achieve a balance of all aspects. Nash, at this point in his life, had already been in constant battle with chaotic elements and is only now slowly beginning to work toward stability. This is, of course, an extreme oversimplification. Still, it begs the question of how outliers such as Nash inform or contradict developmental models such as those proposed by Levinson. Because Nash's life diverts even further from the Levinson model, the Age 50 Transition period is omitted.

The Culminating Life Structure for Middle Adulthood (Age 55–60)

The end stage of middle adulthood is a period of increased stability during which individuals are often able to rejuvenate and enrich their lives (Levinson et al., 1978). This can be a period during which an individual begins to take account of their position in life and explore new directions such as changing their career or changing their personal relationships with others (Levinson, 1996).

Nash's Culminating Life Structure for Middle Adulthood (1984–1989)

For Nash, this was a period when he saw slow remission and recovery from his delusions. Nash himself explained that he intellectually interrogated his delusions, demanding they justify themselves. He claimed this method enabled him to best determine reality from the illusory (Nasar, 1998). In his view, maintaining stability between the real and the delusions required constant surveillance. According to Capps, this position implies that the left hemisphere was in greater control than during the times of greatest mental instability (Capps, 2011). Notably, Nash would later elaborate on this shift in thinking styles and how it changes one's view of both oneself and one's relation to the world:

> So at the present time I seem to be thinking rationally again in the style that is characteristic of scientists. However this is not entirely a matter of joy as if someone returned from physical disability to good physical health. One aspect of this is that rationality of thought imposes a limit on a person's concept of his relation to the cosmos. For example, a non-Zoroastrian could think of Zarathustra as simply a madman who led millions of naive followers to adopt a cult of ritual fire worship. But without his "madness" Zarathustra would necessarily have been only another of the millions or billions of human individuals who have lived and then been forgotten.
>
> *Nash-Jr., 1994*

The Late Adulthood Stage (Age 60–85)

During this late transition, individuals come to terms with their impending retirement and the significant life changes it will bring (Levinson et al., 1978). This is essentially the period during which the individual realizes their professional development is drawing to a close, and they will soon need to find a different organizing structure for themselves.

Nash's Late Adulthood Stage (1989–2015)

Nash's health had improved to the point where he could return to full teaching responsibilities throughout the 1990s. He received the Nobel Prize for Economics despite many in the mathematics community believing he had died years ago (Nasar, 1998). The recognition which came with the award spurred a long period of renewed activity and confidence. While he was not asked to give a Nobel lecture due to concerns over his stability, he did give a short speech at a private party in Princeton. According to Nasar, Nash only had three things to say to the collected group: 1. He hoped the award would improve his credit rating because he really wanted a credit card; 2. he wished that he had won the entire prize for himself because he needed the money very badly; and 3. he felt that game theory was mostly an intellectual interest for academia that the world wishes could have some practical use (Nasar, 1998).

In 1998, Sylvia Nasar's biography was released to immediate critical praise. Notably, Alicia helped Nasar with her research, while Nash himself refused to have any part of the book. His autobiographical essay written for the Nobel Committee explicitly and deliberately omits "details of truly personal type" (Nash-Jr., 1994). He wished for any biographies of his life to strictly focus on his scientific and intellectual achievements. There was even some criticism of Nasar's work as being a "drastic violation of privacy" (Rubinstein citing John Milnor, 1999) by a fellow mathematician. This was in keeping with the recurring trend throughout Nash's life of his wanting only to be known for his genius and contributions to science. When the book was adapted into a film two years later, with the rise of recognition came unwanted attention to several aspects of his life. His treatment of Eleanor and their son, rumors of homosexual behaviors, and anti-Semitic comments made during the height of his illness were all made public knowledge. If having the most private and salacious details of his life made into public gossip bothered him, he never publicly spoke about it. This was possibly due to more significant problems occurring in his family at the time.

Nash and Alicia's son Jonathan had been diagnosed with schizophrenia at age 15, half the age at which his father had been diagnosed. Like his father, Jonathan was extremely talented with mathematics and earned his Ph.D. However, unlike his father, Jonathan did not respond well to contemporary treatments. Despite incredible advances in medication and other treatments, nothing helped him. He was unable to work and had to rely solely upon his parents for everything in his life. Nash took up the role of the supportive father in much the same way his father had done for him. In this late stage of his life, he even made amends with Eleanor and his first son, John David Stier. In 2001, Nash and Alicia remarried. Though divorced, the two had remained together in the intervening years. Alicia's support and care are often cited as a tremendous contributor to Nash's recovery.

On May 23, 2015, John and Alicia Nash were killed in a car accident on the New Jersey Turnpike. They were returning home from a flight from a trip to Norway, where Nash received the Abel Prize in recognition of his work. He was 86, and she was 82.

Conclusion and Limitations

This study applied a simplified version of a method employed by Fouché et al. (2017) to develop a psychobiography of Steve Jobs. Their work intended to find support for the relevance of Levinsonian theory as a means to gain a life cycle understanding of their subject's existence and influence. The aim of this current study was quite different. The life of John Nash, Jr. has been the subject of an exhaustive biography and multiple papers debating aspects of his life, his illness, and his contributions to the field of mathematics. However, there has been no full psychobiography on his life, possibly due to Nash's status as an outlier for many of the approaches that might commonly be used. He was considered a genius. He was an individual with schizophrenia whose symptoms and remission do not correspond to most models of the disorder. His field of study has a reputation for brilliant pioneers who struggled with mental health. He successfully managed the worst of his symptoms by way of applying reasoning to the madness. Any of these factors would prove challenging to develop a study design. Yet, the uniqueness of Nash's life story is precisely why such a project should be attempted.

The findings of this current study first appear to contradict Levinson's theory that the central components of an individual's life have a significant impact on life structure development and the process of individuation. Nash largely preferred to exist within the world of numbers and science, and he explicitly sought to have his life defined by the measure of his work in these areas. However, Nash was primarily motivated by the desire to be recognized and respected for his genius. One of the great ironies of his life is that he eschewed social engagements while at the same time desiring attention from those same people. His career was the most central component of his life. Family became essential in his life but only in the later years. His two romantic relationships were marked by his indifference and hostility toward his partners when their actions interfered with his ambitions. He appeared to recognize later that without the support of his family, he would have perished during the years when his illness was at its most debilitating. He also remained occupationally productive until his death, despite long periods of inactivity due to this illness.

How might someone make sense of Nash's life and work with reference to Levinsonian theory? One could argue that despite its irregularities, Nash did engage in the ongoing individuation process and ultimately became an integrated person. But this argument is undermined by multiple indicators that Nash experienced internal distinctions between himself and his thinking. In a sense, he was fully cognizant that he had to remain separate from aspects of his own consciousness. For this reason, the author included a brief discussion of framing regarding the cognitive aspects of schizophrenia. Is there room in the psychobiographical method for incorporating such an investigation into the modes of reasoning used by the subjects, and how can such models best be integrated with those that focus on experiential dimensions—the life as lived by the subject of the study? The answer to this question is outside the scope of this paper but nevertheless should be addressed.

This study was limited to a method adopted to explore a single theory of adult development. Additionally, most of the information regarding Nash was taken from a single source, Nasar's biography. The author hopes that this study has at least provoked interest in pursuing a more thorough explanation of Nash's psychosocial development and perhaps that his case serves as an illustration of the ways social, personal, and cognitive processes interrelate in making possible even the work of mathematical genius.

QUESTIONS FOR DISCUSSION

1. How can psychobiographical study enhance understanding of mathematical cognition?
2. How can cognitive neuroscience be effectively integrated with development and experiential studies of persons to enhance understanding of creativity and genius?

References

Andreasen, N. C. (1987). Creativity and mental illness: Prevalence rates in writers and their first-degree relatives. *American Journal of Psychiatry, 144*, 1288–1296.

Andreasen, N. C. & Nash, A. (2015). A beautiful love story. *American Journal of Psychiatry, 172*(8), 710–713. doi: 10.1176/appi.ajp.2015.15060709. PMID: 26234595.

Beaty, R. E., Benedek, M., Wilkins, R. W., Jauk, E., Fink, A., Silvia, P. J., Hodges, D. A., Koschutnig, K., & Neubauer, A. C. (2014). Creativity and the default network: A functional connectivity analysis of the creative brain at rest. *Neuropsychologia, 64*, 92–98. https://doi.org/10.1016/j.neuropsychologia.2014.09.019

Bell, E. T. (2008/1937). *Men of mathematics*. New York: Simon & Schuster. (Same text, newer edition).

Brod, J. H. (1997). Creativity and schizotypy. In G. Claridge (Ed.), *Schizotypy: Implications for illness and health* (pp. 274–298). New York: Oxford University Press.

Buckner, R. L. (2013). The cerebellum and cognitive function: 25 years of insight from anatomy and neuroimaging. *Neuron, 80*(3), 807–815. https://doi.org/10.1016/j.neuron.2013.10.044

Burton, D. M. (2010). *The history of mathematics: An introduction* (7th ed.). New York: McGraw-Hill Professional.

Capps, D. (2011). John Nash, game theory, and the schizophrenic brain. *Journal of Religion and Health, 50*(1), 145–162. www.jstor.org/stable/41349773

Dahlberg, K. (2006). 'The individual in the world – The world in the individual': Towards a human science phenomenology that includes the social world. *Indo-Pacific Journal of Phenomenology, 6* [Special Edition: Method in Phenomenology], 1–9. doi: 10.1080/20797222.2006.11433932

de Manzano, Ö, Cervenka, S., Karabanov, A., Farde, L., Ullén, F., & Rustichini, A. (2010). Thinking outside a less intact box: Thalamic dopamine D2 receptor densities are negatively related to psychometric creativity in healthy individuals. *PLoS ONE, 5*(5), e10670.

Dietrich, A. (2014). The mythconception of the mad genius. *Frontiers in Psychology*, https://doi.org/10.3389/fpsyg.2014.00079.

Folsom, D. P., Lebowitz, B. D., Lindamer, L. A., Palmer, B. W., Patterson, T. L., & Jeste, D. V. (2006). Schizophrenia in late life: Emerging issues. *Dialogues in Clinical Neuroscience, 8*(1), 45–52. https://doi.org/10.31887/DCNS.2006.8.1/dfolsom

Fouché, P., du Plessis, R., & van Niekerk, R. (2017) Levinsonian seasons in the life of Steve Jobs: A psychobiographical case study. *Indo-Pacific Journal of Phenomenology, 17*(1), 1–18. doi:10.1080/20797222.2017.1331970

Friston, K. (1998). The disconnection hypothesis. *Schizophrenia Research, 30*, 115–125.

Garrity, A. Pearlson, G., McKiernan, K., Lloyd, D., Kiehl, K., & Calhoun, V. (2007) Aberrant "default mode" functional connectivity in schizophrenia. *American Journal of Psychiatry, 164*, 450–457.

Greenwood, T. A. (2016). Positive traits in the bipolar spectrum: The space between madness and genius. *Molecular Neuropsychiatry, 2*(4), 198–212. https://doi.org/10.1159/000452416

Gruber, H. E., & Bödeker, K. (2005). *Creativity, psychology and the history of Science.* Heidelberg: Springer Netherlands.

Hare, E. (1987). Creativity and mental illness. *British Medical Journal, 295*, 1587–1589.

Hausner, M., Nash, J. F., Shapley, L. S., & Shubik, M. (1964). So long sucker, a four-person game. In M. Shubik (ed.) *Game theory and related approaches to social behavior* (p. 90–93). New York: John Wiley & Sons, Inc.

Howard, R. (2001). *A beautiful mind.* United States: Universal Pictures.

Kang, S. H., & Son, C. N. (2017). 심리전기적 정보(Psychobiographical Information)가 정신질환자에 대한 태도변화에 미치는 효과. [The effect of psychobiographical information on attitude change toward mentally ill people.] *Journal of Digital Convergence, 15*, 451–457.

Kant, I., & Meredith, J. (2007). *Kant's critique of judgement.* London: Oxford University Press.

Karlsson, J. (2004). Psychosis and academic performance. *British Journal of Psychiatry, 184*, 327–329.

Kaufman, S. B. & Paul, E. S. (2014). Creativity and schizophrenia spectrum disorders across the arts and sciences. *Frontiers in Psychology, 5*, 1145. https://doi.org/10.3389/fpsyg.2014.01145

Kuhn, H. W., & Nasar, S. (2007). *The essential John Nash* (pp. 5–6). Princeton, NJ: Princeton University Press.

Levinson, D. J. (1986). A conception of adult development. *American Psychologist, 41*(1), 3–13. doi:10.1037/0003- 066X.41.1.3

Levinson, D. J. (1996). *The seasons of a woman's life.* Washington: Ballantine Books.

Levinson, D. J., Darrow, C. N., Klein, E. B., Levinson, M. H., & McKee, B. (1978). *The seasons of a man's life.* New York: Ballantine Books.

Ludwig, A. M. (1995). *The price of greatness: Resolving the creativity and madness controversy* (pp. 1–12). London: Guilford Press.

Narahari, Y. (2016). Beautiful results from a beautiful mind. *Resonance, 21,* 770–801.

Nasar, S. (1998). *A beautiful mind.* New York: Simon & Schuster.

Nash-Jr., J. (1994) John F. Nash Jr. autobiography. In T. Frängsmyr (Ed.), *Les Prix Nobel/ Nobel Lectures/The Nobel Prizes.* Stockholm: Almqvist & Wiksell International.

Perina, K. (2017). The mad genius mystery. *Psychology Today.* Retrieved May 15, 2022, from www.psychologytoday.com/us/articles/201707/the-mad-genius-mystery

Ponterotto, J. & Reynolds, J. (2013). The "genius" and "madness" of Bobby Fischer: His life from three psychobiographical lenses. *Review of General Psychology,* Advance online publication. doi:10.1037/a0033246

Ponterotto, J. & Reynolds, J. (2017). Ethical and legal considerations in psychobiography. *American Psychologist, 72*(5), 446–458.

Raichle, M. E., MacLeod, A. M., Snyder, A. Z., Powers, W. J., Gusnard, D. A., & Shulman, G. L. (2001) A default mode of brain function. *Proceedings of National Academy of Science, USA, 98,* 676–682.

Rubinstein, A. (1999). A review on: 'A beautiful mind: A biography of John Forbes Nash, Jr.' by Sylvia Nasar. *Ariel Rubinstein Home Page: Game Theory.* https://arielrubinstein.tau. ac.il/

Shorter, E. (2015, May 27). A beautiful mind: What did John Nash really have? *Psychology Today.* www.psychologytoday.com/us/blog/how-everyone-became-depressed/201 505/beautiful-mind-what-did-john-nash-really-have

Suellentrop, C. (2001, December 21). A *beautiful mind's* John Nash is less complex than the real one. *Slate Magazine.* https://slate.com/culture/2001/12/a-beautiful-mind-s-john-nash-is-less-complex-than-the-real-one.html

van Genechten, D. (2009). A psychobiographical study of Helen Keller. [Doctoral dissertation, Nelson Mandela Metropolitan University.] https://core.ac.uk/download/pdf/ 145049785.pdf

Whitfield-Gabrieli, S., Thermenos, H. W., Milanovic, S., Tsuang, M. T., Faraone, S. V., McCarley, R. W., Shenton, M. E., Green, A. I., Nieto-Castanon, A., LaViolette, P., Wojcik, J., Gabrieli, J. D., & Seidman, L. J. (2009). Hyperactivity and hyperconnectivity of the default network in schizophrenia and in first-degree relatives of persons with schizophrenia. *Proceedings of the National Academy of Sciences, 106*(4), 1279–1284.

6

AN INTERPRETATION OF FRANZ BOAS' CONTRIBUTIONS TO ANTHROPOLOGY AND SCIENTIFIC ANTIRACISM

Merging a Psychology of Science with Anthropology and Feminist Theory to Discuss the Human Influence in Science

Georgia F. Crowe

Personal Preamble

I was introduced to Franz Boas as an undergraduate student with a double major in psychology and anthropology. It is an overarching theme of anthropology in the United States that an undergraduate, and sometimes even a graduate degree, will be a four-field pursuit. This means that the coursework and other requirements are fairly evenly distributed between four dominant areas: cultural anthropology, linguistic anthropology, physical anthropology, and archaeology. This approach of American anthropology is itself "Boasian." However, students often take up one of these areas as a central focus depending on their academic and vocational interests. My own special interest was in cultural anthropology, which places a great deal of focus on the lives of people within their social world and the various institutions and practices that we understand as culture. Within this subfield in particular, Boas is an extremely prominent figure.

More personally, the anthropology professor that has had the most influence on me, both in my personal life as well as in academia, is dominantly Boasian in her own teaching style and role within the field. This has undoubtedly had an influence on how I see Boas as a prominent figure not only in the field as a whole but in my own academic experience. Specifically, it is Boas' belief that we must examine all areas of a culture in order to understand it most fully, must rely on or gain a native, or emic, perspective, and operate with cultural relativism, that has been most influential to how I conduct my own research. This has helped me to

DOI: 10.4324/9781003276692-8

think more critically and holistically about various elements that might play a role in the projects I take up. Boas' work has encouraged me to focus on one's subjective experience of a phenomenon.

This is a view that plays a much bigger role in my life than in just academia: these are also perspectives that have drastically changed the way I view people and cultures throughout my personal life also. Cultural relativism as well as reliance on the emic perspective have been huge influences in the way in which I approach learning more about the world around me. These have been tools that, since being introduced to them, have made me more understanding, more empathetic, and a better ally to communities other than my own. They have encouraged me to grow in my view of the world, broaden my belief system, and become more inclusive. As I am sure is the case for many young adults entering higher education, I have learned much about the world and how to have a more open mind about the way in which it works, how people within it live and believe, and the ways in which our differences make us unique and equally valuable. Boas specifically, as well as the professor and courses through which I have learned about him, have aided in this process the most. Overall, Boas was someone that was an incredibly important figure in shaping how I do research now as well as who I have grown to become as a person, and for the sake of being an advocate for interdisciplinary studies, this is a particularly useful opportunity to be able to bring together some of my passions in other fields by presenting a case study of Boas through the lens of psychology of science which reflects on additional contributions from feminist theory.

Introduction

Ideas and Contributions Part One

As previously mentioned, Boas' first notable contribution to the field is that he is the most prominent figure associated with creating a four-field approach in American anthropology (King, 2019). His own theory and methods are methodology informed by work from each of the four subfields, but he is most often regarded as an ethnographer and a cultural anthropologist because of his focus on fieldwork. His belief that only persons within a particular culture could teach academics about that culture was hugely influential in anthropology, and his own methods furthered a framework for understanding the fundamentals of fieldwork. However, we can see the four-field approach across the body of his own research. His study of immigrants at Ellis Island largely implicates him as a physical anthropologist because he was most concerned with physical measurements of humans and their characteristics. The museum work he did put him in the role of archeologist. His interest in the language of the cultures he studied cast him as a linguist. His earliest ethnographic study allowed him to develop a methodology that helped cultural anthropology develop as a science. In combination, the simple fact that Boas utilized so many diverse approaches to anthropology established American anthropology as a four-field discipline.

Important to note, though, is that Boas' personal and educational background established the groundwork that enabled such a foundational contribution. According to Darnell, "[t]he distinctive features of the Americanist tradition [of anthropology] arose from the German romanticism that Boas imported to North America by virtue of his European education" (Darnell, 2016, p. 3). This tradition, in turn, influenced the way Boas understood culture "as a symbolic form, a construction in people's heads rather than a thing accessible to direct observation," and as something to be understood holistically (Darnell, 2016, p. 3). Yet we must also recognize the innovation of his approach, the grounding of "the new anthropology" in "first-hand fieldwork," yielding data that "could be mined for linguistic, ethnological and/or psychological insights," requiring anthropologists to immerse themselves in communities, gain trust, and form relationships (Darnell, 2016, p.3). Before Boas, anthropology was mostly conducted unsystematically, and in the service of philosophy and theory. Boas, and others who were influenced by his work, imbued it with a more well-rounded and better-informed methodology, giving the anthropologist a more active, applied role in the field.

The second and third most important contributions Boas made to anthropology are the concept of cultural relativism and the theory of historical particularism. According to King (2019), Boas and his students "called themselves cultural anthropologists—a term they invented—and they named their animating theory cultural relativity, now often known as cultural relativism" (King, 2019, p. 8). Simply put, cultural relativism is the view that no culture is more or less valuable, morally sound, or "correct" in the way of living than another, and while, of course there can be practices in a culture that are ethically unsound, we cannot make such judgments until we fully understand the practices from the native people's point of view. Boas taught that anthropologists cannot simply observe culture, which was the dominant focus of museums and anthropology at the time. He maintained that culture is what is experienced and perceived by the individuals situated within it. Therefore, someone could communicate their culture to scientists, who could also learn about a culture by being situated within it to gain an insider's perspective, but one could not study it in the same way one studies the natural sciences. Importantly, this way of viewing culture has implications beyond the specific practices of anthropology. Boas believed that:

> in order to live intelligently in the world, we should view the lives of others through an empathic lens. We ought to suspend our judgment about other ways of seeing social reality until we really understand them, and in turn we should look at our own society with the same dispassion and skepticism with which we study far-flung peoples.
>
> *King, 2019, p. 8–9*

The overarching implication of this idea is that there is no culture that operates as a standard or blueprint for how other cultures should operate. This was a radically different view than that of the dominant paradigm of anthropology in Boas' time:

> A science that seemed to prove that humanity had unbridgeable divisions had to be countered by a science that showed it didn't. By making Americans in particular see themselves as slightly strange… Boas and his circle took a gargantuan step toward seeing the rest of the world as slightly more familiar.
>
> *King, 2019, p. 13*

What has come to be labeled "historical particularism" by later anthropologists is based on the argument Boas made that additionally contributed to way anthropology understands culture. For example: "Boas argued that race, language and culture were independent variables subject to vicissitudes of contact, migration, and environment, formulating a paradigm retrospectively labeled as historical particularism" (Darnell, 2016, p. 3). What this ultimately amounts to is that cultures are regarded as the product of the many particularities of the time and place in which they exist. It is the view that these variables change over time; moreover, "there is no more fundamental reality in the social world than the one that humans themselves create" (King, 2019, p. 9). For Boas, this was not simply a way of broadening and strengthening one's field of study; it was a matter of utmost importance for the very safety, equality, and well-being of the peoples of the world, and it was related to his own personal and sociocultural experiences within such a system.

After this overview of Boas' primary contributions to anthropology, we now turn to an examination of some of the details of his life through a brief biographical sketch. This will help to establish a basis upon which the personal, social, and cognitive contexts of these contributions can be better analyzed. The biography is based primarily on two sources: Darnell (2016) and King (2019), the limitations of which will be discussed later in the chapter.

Biography

Franz Uri Boas was born on 9 July 1858 (Baker, 1998–1999). His family lived in Minden, a small town in Prussia, which would later become northern Germany (King, 2019). His family is described as "bürgerlich—urban, educated, freethinking, bourgeois—[which] was as much a defining feature of life as being members of a minority faith" (King, 2019, p. 15). The decade leading up to Boas' birth held a period of armed uprisings and public demonstrations known as the "Springtime of the Peoples." This movement was led by workers and academics alike with the aim of reform, social justice, and more unification across Germany (King, 2019). During the following season, however, monarchs and prime ministers were able to regain their power, and the revolution ultimately failed. While this was an event that occurred years before Boas' birth, his mother, Sophie, instilled in Boas many "pedagogical and political ideals of the failed revolution of 1848" (Darnell, 2016, p. 2). Despite growing up with so many influential ideas around him, a young Boas was resistant to attending university because "his guiding principle was that of many talented teenagers: to try to arrange things so that he would not become 'unknown and unregarded'" (King, 2019, p. 17).

However, in 1877, he did enroll in Heidelberg University, where he was influenced by Immanuel Kant in particular. King (2019) provides an excellent, concise summary of some of the views that might have influenced Boas: "rather than be skeptical about everything we claim to perceive, the surest route to true knowledge was to turn our attention toward our perceptions themselves" (King, 2019, pp. 18–19), "we are all, by definition, experts in our own experience" (King, 2019, p. 19), and "the way to understand something about the world was to steer a course between a belief in the universal power of reason and an unbending skepticism about our ability to know anything at all" (King, 2019, p. 19). Boas was also influenced by the important emphasis for younger professors on "the relationship between physical reality and human perception" (King, 2019, p. 19). This is particularly present in Boas' later work with race which looks at societal perceptions of race compared to the physical reality of our characteristics. However, at this time, Boas put these ideas to work in a dissertation that allowed him to perform his own research which was required for a Ph.D. (King, 2019). King (2019) describes this work as "all inexpert and improvisational but enough to gain a grudging pass from the examiners," resulting in a Doctor of Philosophy of physics in 1881 (p. 19). Boas' study had focused on light characteristics in its relation to water, but he had more of an interest in how we as perceivers encounter the physical world, for example, "[figuring] out not what the natural world *does* but how we determine for ourselves what we *think* it is doing" (King, 2019, p. 20). This pursuit needs to begin, he believed, an effort to understand the way others who are much different from us might view the world. For Boas personally, that meant "getting as far away as possible from familiar places such as Minden and Kiel" (p. 20).

It was at this point in his life that Boas was able to convince his father to fund an expedition to the Arctic in which he conducted a study of the hunting and migration of Inuit peoples living on Baffin Island; this was his first experience with studying culture and the inception of his interests in anthropology and fieldwork. It was driven by what King describes as a "call to duty, instilled in every German schoolchild of means to add to Germany's national greatness by reaching the ends of the earth before other nations got there" (King, 2019, p, 21). He even acquired a position writing a series of articles on his experience with a newspaper. He lived among the Inuit at Baffin Island until 1884 when he first attempted to find employment in the United States with many failed connections made by his uncle living in New York. However, each of these attempts lead to a dead end. Although he was able to secure a deal to publish some of his research from Baffin

Island, this did not feel like a success for Boas, despite being invited to present in Washington:

> This was at least something to show for his trip to Washington, but even then, Boas worried that the money ... offered would not be sufficient to cover the project's full cost. Maps would have to be drawn and etchings made. He would also need help with his English. His command of the language might have impressed [companions throughout his study], but it did not move his American hosts. He found himself unable to follow the discussion during a meeting of one of Washington's scholarly societies, and a secretary was forced to read Boas' paper aloud while he looked on in silence. He soon returned to New York, depressed and embarrassed. Two lectures that he was invited to give at Columbia College, arranged by Uncle Jacobi, proved yet another linguistic disaster.
>
> *King, 2019, p. 36*

Boas was forced to return to Germany in 1885, seemingly plagued by insecurity and a depressive episode. For a period, Boas continued his education, research, and travels until he was able to immigrate to the United States, once again with the aid of his uncle. Boas worked for a time at *Science* magazine, made connections with Granville Stanley Hall who offered him a teaching position with Clark University, and eventually served as a staff member of the Chicago World Fair in which he was first exposed to large-scale studies of anthropometrics, measurements of the human body, which drastically shifted his career.

The Personal Dimension

Having now examined an overview of both Boas' contributions to anthropology and then some important biographical details, we can begin to reflect on how his life and work are interrelated, beginning with what we are here calling the personal dimension or personal context of his many contributions. We should note, though, how the personal and social dimensions of his work are intimately interrelated for Boas.

The personal dimension may consist of personality characteristics or disposition, values, beliefs and commitments, and similar factors. Of note are the many connections between the personality Boas is often described as having and Grosul and Feist's (2014) analysis of personality traits of creative artists and creative scientists. The meta-analysis in that study showed creative personalities demonstrate trends in openness to experience and impulsivity (Grosul and Feist, 2014). Additionally, where there was the comparison between creative artists and scientists, Boas seems to have traits from both groups. Creative artists tend to be more unstable, imaginative, and impulsive whereas creative scientists tend to be more arrogant and ambitious (Grosul and Feist, 2013). Boas is described as showing all of these characteristics. Darnell (2016) suggests that "he was not a tactful or patient man" (p. 2). His abrasiveness in his professional work is a recurring theme throughout his biography (King, 2019).

The influence of his environment and life circumstances on Boas' temperament is important to note. King describes him as having a combative social structure in his university days, one in which the university students "divided themselves almost instantly into associations of friends and confidants whose only real duty was to police the boundaries of the very associations they had made" (King, 2019, p. 17). Such ties to duty led to frequent skirmishes and dueling among the students, eventually leaving Boas with scars from a saber across his face. In later life Boas seemed embittered by the appearance of tumor on one of his salivary glands. Despite the successful removal of the growth, Boas suffered long-term effects such as facial drooping and blurred vision from a severed nerve during the surgical procedure. King notes that Boas saw this as a "kind of death sentence," and linked it to "his blistering critiques of the [First World W]ar: a dying man had nothing to lose" (King, 2019, p. 115). Despite his physical challenges, Boas seems to have remained resilient, energetic, and socially confident. King (2019) provides the description:

> What Boas did have was energy. He was a talker, someone with little compunction about contacting people he didn't know or showing up at their offices with a long list of plans for expedition or revolutionary hypothesis he was burning to describe. He might lead off with a story about how his facial scars had been sustained in a polar bear attack, leaving his listener to wonder whether he was joking.
>
> *King, 2019, p. 31*

Social Dimensions

The Experience of "Otherness"

Social processes embed scientific achievement at both local or "micro" and broader, more general "macro" levels. For Boas, important social processes at the macro level of Boas' introduction to American culture included a broad pattern of dehumanizing people of color and a widespread belief that they were not equal to white westerners. In the United States, immigrants were seen as inferior to natural born citizens. Western civilization was seen as the height of culture, a belief that traces back to colonialism and continued to escalate throughout the Civil War and much later; however, even those who immigrated to the United States were seen as harming society rather than joining, strengthening, and diversifying it. Racial categorizations were particularly pervasive throughout society which can be reflected by "censuses divided society into clear and exclusive racial types …Your proper category was so obvious that it was not what *you* said it was but what *someone else*, the census enumerator usually a White man—said it was" (King, 2019, p. 5). These categorizations shaped criteria in academics, political policy, ideas of health, and societal perception (King, 2019). It was an overarching belief throughout the general society, academia, and politics that sang the praises of

> the robust colonizer over the benighted native. Differences in physical appearance, customs, and language were reflections of a deeper, innate otherness. Progressives, too, accepted these ideas, adding only that it was possible, with enough missionaries, teachers, and physicians on hang, to eradicate primitive and unnatural practices and replace them with enlightened ways.
>
> *King, 2019, p. 6*

The dangerous rhetoric of academia and politics were not just philosophical musings on the nature of humanity; they were harmful ideologies that permeated throughout society, resulting in very real harm. This is a trend that continues even today as these ideas remain present within our current society.

While Boas was not a person of color and obviously could not know that experience, he faced marginalization for being Jewish, for being German-born, and as an immigrant to the United States: "His Judaism, never mind his reportedly almost unintelligible German accent, rendered him an outsider at best" (Darnell, 2016, p. 3). His studies were very clearly linked to experiences he had and things he witnessed happening to others that did not meet the expectations of the western "ideal." For example, during the more turbulent years of Boas' career, he was hired at Clark University by G. Stanley Hall, but his time there was fraught with controversy, with his project of obtaining measurements of local school children leading to "public outcry," which in Darnell's view was fueled by his being "a man, a foreigner and a Jew after all!" (Darnell, 2016, p. 2). His uncle aided him in finding new opportunities, and though jobs were often denied him, he was ultimately hired at Columbia University and the American Museum of Natural History in 1897 (Darnell, 2016). As an immigrant with multiple intersecting marginalized identities, Boas was fortunate to have familial ties in the states that were able to aid him in launching his career regardless of the turbulence of establishing his name in the field.

Intellectual Zeitgeist

It is important to consider the intellectual zeitgeist of Boas' era which is one rooted within this very same much deeper history. During and after the Civil War, a popular idea propagated among American scientists was polygenesis, referring to the idea that the human species had multiple points of origin (Baker, 1998–1999). Notable is that these views persisted even following the publication of Darwin's *On the Origin of Species by Means of Natural Selection* which provided data proving many variations were possible within one species based on environmental factors (Baker, 1998–1999). Such beliefs may be traced at least to 1775 when the first official classification system of race was developed, based on the supposedly different origins of different races. They continued into the 1880s with the inception of the term *eugenics* and development of anthropometry, a means of measuring physical characteristics of the body that were believed to be predictors of health, intelligence, and even social deviancy (Baker, 1998–1999; King, 2019). In anthropology, the impact of these ideas was felt as ethnologists who were Social Darwinists promoted the idea that cultures, and their peoples, could be categorized in comparison to one another to show an "evolution" with the perceived highest classification of culture being that of western civilization (Baker, 1998–1999). Importantly, these ideas were not simply confined to the ivory tower, with theorists and scholars discussing race and culture in hypothetical terms. These were stances that were shaping the ways in which the country developed and they were frequently taken up by government representatives who utilized them to sway public opinion. Thus, there were very serious connections between what was being done in this area of academia and what was happening to people of color as well as immigrants. The consensus among scientists during this time was that white Americans were the height of human culture, and that while assimilation of other groups was seen as bettering those groups, there was little expectation that they could meet the standards of civilization that had been set by colonizers. The legacy of such a tradition is the ethnocentric and racist theories that have been produced throughout the history of studying other cultures which is still being combatted and healed from today.

Cognitive Processes

When considering classifications of culture, Boas thought the only way one could even begin to develop any categorization was through the inductive method—and by examining data in depth and suspending theorizing until you gathered as much data as possible. However, he saw the ideas that were emerging in the field as purely deductive, and biased, in which theorists were developing a model and forcing what they perceived into what they had already framed themselves to believe to confirm their own prejudices. Studies of culture and race had primarily developed through deductive methods:

> The researcher is required to take a hypothesis or a theory and test it indirectly by deriving from it one or more observational predictions. These predictions are amenable to direct empirical test. If the predictions are borne out by the data, then that result is taken as a conforming instance of the theory in question.
>
> *Haig, 2005, p. 372*

Singer (1971) explains that once one forms a hypothesis, it is hard to be swayed by disconfirmation, so, for Boas, what he saw being conducted in the sciences at the time was merely confirmation of long-standing prejudices that were lacking in data but religiously held in academia, fueling racism, xenophobia, and ethnocentrism in politics and the general population. Boas believed that what was needed in studies of race and culture was inductive reasoning, namely the suspension of judgments until thorough data had been collected in order to then draw accurate conclusions to develop an informed discussion of race. Haig (2005) provides a useful definition of this process as

> the scientist is typically portrayed as reasoning inductively by enumeration from secure observation statements about singular events to laws or theories in accordance with some governing principle of inductive reasoning. Sound inductive reasoning is held to create and justify theories simultaneously, so that there is no need for subsequent empirical testing.
>
> *p. 372*

Furthermore, commitment to confirming one's preexisting belief can result in fraudulent work and, as we see in the broader field at the time, to acceptance of what is not present within the data, or the use of forced data: "it is quite important to keep in mind that while commitment and discovering are proposed as necessary experiences for the scientist, they do not imply that knowledge thus created is necessarily correct or novel" (Carré, 2019, p. 20). We may also refer to such work as empirical, meaning that it "defends experience rather than ideas as the source of knowledge and is thus contrasted with rationalism" (Harding, 1991, p. 112). This is not to say that there were no ideas involved, only that the reasoning processes is grounded and verified in first-hand experience.

The cognitive dimension, then, entails commitment on the part of the scientist as much as it does a pattern of reasoning. It is therefore very closely connected to the personal and social dimensions of the scientific activity. Carré (2019) notes that "the scientist's preference for certain data, methods, and theories is ultimately rooted in a series of life experiences that led him or her to be certain that these are the most accurate and pertinent for the inquiry at hand" of which the most essential are experiences of "formal education and scientific specialization" (Carré, 2019, p. 18). Boas founded his mode of thinking in first wondering about our experience and then finding ways to explain it, focused his ways of seeing the world through such experiential means rather than merely the physical, and to be generally curious about the world around him. This came at least in part through his German educational tradition, his access to which was enabled by his class and status. However, what Boas saw unfolding for future students as well as for people of the United States was a concerning national mindset, as well as an education system that could become reliant on unhelpful ways of viewing race, and that would be nearly impossible to undo once firmly entrenched.

The Personal, Social, and Cognitive: Boas and Race

Earlier in this chapter I noted three lasting, foundational contributions Boas made to the field of anthropology. Here it seems important to acknowledge a final important contribution of Boas' scientific reasoning, grounded as it was in inductive processes. But this is one that spread beyond the field of anthropology and even beyond the social sciences into society more broadly. This was Boas' impact on the concept of race. Throughout his career he used his empirically grounded research to outright disprove and discredit the many racist and ethnocentric views held within science, academia, and society. This was especially impactful because he is credited with essentially defining the direction anthropology was to take. Boas took a more prominent leadership role in the field, orienting anthropology away from ethnocentric and hierarchical theories of race and toward the focus that all races and ethnicities contained their own context-specific cultures.

These cultures, he argued, were particular to geographic areas, local histories, and traditions. "Furthermore, one could not project a value of higher or lower on these cultures—cultures were relative" (Baker, 1998–1999, p. 94). The impact such a view had on the dominant paradigm can be understood as follows:

> The implications of the idea that we make our own agreed-upon truths were profound. It undermined the claim that social development is linear, running from allegedly primitive society to so-called civilized ones. It called into question some of the building blocks of political and social order, from the belief in the obviousness of race to the conviction that gender and sex are simply the same thing. The concept of race, Boas believed, should be seen as a social reality, not a biological one—no different from the other deeply felt, human-made dividing lines, from caste to tribe to sect, that snake through societies around the world.
>
> *King, 2019, p. 9*

Baker (1998–1999) reports that

> as early as 1887 Boas began to combat scientific racism by challenging museum organizers' representations of other cultures. He argued that arranging artifacts into categories depicting degrees of savagery, barbarism, or civilization employed a fraudulent logic "not founded on the phenomenon, but in the mind of the student".
>
> *p. 94*

This is a direct challenge to concepts as Social Darwinism, a trajectory for the evolution of culture that results in a Western example of civilization, as well as shows the need for data, not just theory, in developing such schematic arguments. However, it is Boas' (2011/1911) study *Changes in the Bodily Form of Descendants of*

Immigrants, which made the most impactful, scientifically grounded push for anti-racism in the sciences.

As previously discussed, the ideas of eugenics and attempts at making a hier-archical classification of race and ethnicity were not simply academic theories but were being taken up by policy makers and the public at large to create and enforce oppressive laws and regulations. Academia's attempts to measure and identify race as a marker for success, value, and social influence were not simply theorized ideas to be discussed in the classroom, which would be harmful enough; they were readily taken up by politicians and the general population to defend racist policy and actions. An example of this is the 1907 United States' commission to study the influx of immigration, the attempt of which was to prove that other countries were intentionally sending criminals into the United States to weaken its society. This is a view that was held, and arguably continues to be held, by public figures as well as the general public (King, 2019). In 1908 Boas was hired by the commission to collect data and prepare a report from a study utilizing anthropometry to show that deviant behavior could be linked to one's physical characteristics (King, 2019). The idea was that the study would then be implemented as policy to limit immi-gration. It was believed that the data would show consistent measurements across time and across peoples which would then prove individuals not being physically changed through living in America. In turn, the intention was to provide support for the idea of multiple sources of origin for the species because of the unyielding genetic consistency among groups of color that are different from the consisten-cies of White people.

Boas enthusiastically accepted and responded that he could assure the commission that the results would be influential (King, 2019). The results, Boas already knew from previous work in anthropometrics, would be exactly what would be needed to prove not only academia wrong about their uninformed theories on multiple origins of the species but would also be able to officially discredit the immigration commission's project using its own study! And that's exactly what he did. In 1911, after three years of research, the results of the study were published. On the second page Boas states,

> The investigation has shown much more than was anticipated. There are not only changes in the rate of development of immigrants, but there is also a far-reaching change in type—a change that cannot be ascribed to selection or mixture, but which can only be explained as due directly to the influence of environment.
>
> *Boas, 2011/1911, p.2*

He continues that:

> the bodily traits which have been observed to undergo a change under American environment belong to those characteristics of the human body

which are considered the most stable. We are therefore compelled to draw the conclusion that if these traits change under the influence of environment, presumably none of the characteristics of the human types that come to America remain stable. The adaptability of the immigrant seems to be very much greater than we had a right to suppose before our investigations were instituted.

Boas, 2011/1911, p. 2

The study not only challenged the belief that different races had different sources of origin but demonstrated that the physical characteristic differences that were present among individuals were based on where they lived, lending support to the idea that the body must adapt to the environment.

We now understand race more clearly as a social phenomenon and construct, and from our contemporary point of view Boas' view of race is not without its criticisms. These come particularly from critical studies which suggest that "the attempt to expunge race from social sciences by assigning it to biology, as Boas and his students did, helped legitimate the scientific study of race, thereby fueling the machine of scientific racism" (Lewis, 2001, p. 448). However, Boas' study utilized the very tools and methods that the dominant paradigm was wielding as a weapon of scientific racism to develop a scientific *antiracism* that took some of the first steps in developing the understanding of race we have today. His study of race had lasting impact on the field and society, and it is a study motivated by his personal experiences and values, his view of people and cultures, but conducted with scientific rigor.

Limitations

There is little publicly available quality source material on Boas from which to analyze the psychological contexts of his scientific accomplishment. Darnell (2016) voices such difficulties in saying that "[m]ost of the writing about Boas is by anthropologists and unsurprisingly deals primarily with his career as an anthropologist, although several biographers have emphasized his activism, especially antiracism in North America and critique of Nazi Germany" (p. 3). Lewis (2001) expresses similar frustration:

> It is difficult today to realize the extent of Franz Boas' influence as a scholar, as an institution-builder, and as a public intellectual, because the scope of his work was so enormous, and his impact was so widespread. In the absence of any complete biography, we must depend upon widely scattered articles and chapters.
>
> *p. 450*

A great deal of the material collected on Boas for this biographical sketch and analysis comes from King (2019), as this work is an in-depth biography. While there are additional scholarly articles that serve as great supplementary material, they are still limited in their length and detail. I noticed in particular that King (2019) and others that focused on Boas' biography and contributions to anthropology do not discuss Boas' study of immigrants in a fine-grained way. There is little direct attention to Boas' methods, personal notes about the time in his life in which the study was conducted, or even journals one might expect an anthropologist such as Boas to have kept during conducting fieldwork. During my initial research for a psychological case study of his science, I was overjoyed when I learned Boas had published more than five hundred pages of handwritten data from his famous Ellis Island study. However, upon seeking out the journal, I was not expecting it to contain only charts of the measurements without any additional notes, commentary, or discussion. The published study itself did not provide much more helpful material either. *Changes in the Bodily Form of Descendants of Immigrants* (Boas, 2011/1911) is an investigation conducted by Boas through the United States Immigration Commission, so it is unclear what such an undertaking might have entailed. One might expect Boas to have included much more explanation to accompany his data, given the radically impactful results presented in the data of the study, the implications of his findings, not to mention his own methodological commitments. However, Boas' presence feels relatively small compared to the way in which biographical works about his life frame his criticisms of dominant views of race at the time, as well as his academic activism. There is primarily a focus on the quantitative aspects of the research, relying heavily on charts and figures with little of his own writing. Thus, this analysis is not only an interpretation of a scientist, but an interpretation of historic accounts of a scientist. The present work did not rely on one interpretation, but on layered interpretations in which much may be misconstrued, romanticized, or even unintentionally falsified. That is, the presentation of the biographical information and the analysis of the social and personal contexts in which it unfolded are both interpretive acts on my part, using a particular lens for the purposes of this project. But it is important to acknowledge that this is still useful in furthering our understanding of scientists as people and the influences of the personal, cognitive, and social processes embedding their work and accomplishments.

Another limitation in this case study is that of the political, academic, and personal relationships in Boas' later career. The issue of political and academic relationships arises from Baker's (1998–1999) article which highlights the important contributions of Black intellectuals in putting such ideas to work: "Without the wider social and political efforts of [W.E.B.] Du Bois, the NAACP, and scholars at Howard University, Boas' contributions to the changing signification of race would have been limited to the academy" (p. 95). Baker (1998–1999) notes further that "Boas' reputation began to grow, and he became viewed by African-American leaders as an ally in the struggle for racial equality because of his antiracist research and theories" (p. 95). The article also very briefly touches on the social and political climate of a clash in views between prominent figures such as W.E.B. Du Bois and Booker T. Washington, mentioning that Boas formed a stronger alliance with Du Bois, which associated his work with the "radical integrationist" side of the controversy and also the NAACP. In turn, according to Baked, these alliances "alienated him from the accommodationist wing of the movement led by Washington and financed by Andrew Carnegie. It, in effect, cut off Boas from possible funding from Carnegie" (Baker, 1998–1999, p. 95). This may have been relevant to the climate in which Boas' work came to fruition, leading to a paucity of information regarding connections between Boas' work and the application of his findings.

Boas' career as a teacher receives scant attention in the literature, largely overlooking important relationships with his colleagues after the earlier stages of his career as well as with students apart from a select few names such as Margaret Mead, Ruth Benedict, and other heavily influential figures who continued their careers within anthropology. King (2019) does a particularly thorough job discussing Boas' legacy via prominent figures within the field. However, although covered in more detail within the pages of King's (2019) lengthier biographical work, names as widely known as Zora Neale Hurston are seldom mentioned by other sources in the context of their time as students of Boas. For example, Darnell summarizes his teaching legacy by noting that, "Boas rapidly proved himself a skilled institutional leader who reoriented and reenergized the discipline, producing a talented first generation of students [...]" (Darnell, 2016, p. 3). However, relatively few students are discussed beyond being a student of Boas. While it might be initially understandable that prominent anthropological figures are often considered within the context of their relationship to Boas, there are others who influenced *him* and thus the field of anthropology. As one example of a professional influence, Claude Levi Strauss, a colleague, was close enough to Boas to be present at his death, yet their relationship is rarely discussed. Instead, as noted, Boas is most frequently associated with his student prodigies. Assuming Boas had some influence on these colleagues, and they on him, a psychology of science should be interested *and able* to explore these relationships further.

Finally, Boas' wife, Marie, seems generally to be regarded as an afterthought in the many works on Boas (including the present work), as do their children and the rest of his family once Boas established himself in a more stable position in the United States. Like his professional relationships, Boas' personal relationships, especially those within his familial ties, seem relatively unnoticed outside of the context of beginning his career. The field of psychology of science aims to understand the personal, cognitive, and social processes that might influence the work of a scientist as a person, not simply as a research-producing machine. While there is certainly the chance for additional bibliographic work where source material is available, research on such details of a scientist's work often simply does not exist because it is not adequately recorded. This alone should be an eye-opening revelation for science and technology studies, for academics who do not document their personal lives as thoroughly as their research, and for academia as a whole, that there is a need for a thoroughly supported field of psychology of science.

Conclusion

Personal Resonance

Boas is an important name in the social sciences, particularly anthropology, and while I learned about him from many courses leading up to this assignment, I was unaware of his feelings of rejection and inadequacy concerning his place within the field. He clearly made groundbreaking, empathetic, and needed contributions to the field as a whole, and yet he experienced much failure before accomplishing the things he did. He was very much up against the dominant paradigm of his time, which was not only a constant battle to dispute scientifically, but also something which he had to work to unlearn for himself. For much of his career, even after periods of growth that chipped away at some of these ideas, Boas was a product of the culture of academia in his era. While one's own historical positioning does not excuse their work, it is important to recognize such shortcomings within their respective contexts. Such will surely be the case for this analysis just as it was for Boas'.

There were times he saw individuals and culture through an ethnocentric lens just as those he pushed back against did. This chapter is by no means a praising of Boas' perfection as it does not exist. The process of becoming the scientist he was ultimately to become was long, messy, and nonlinear, as King (2019) makes especially clear. Even simply the focus of what he studied is an example of this: Boas began as a physicist, then became a naturalist, performed quasi-museum curation, became a professor, and throughout worked as an ethnologist and anthropologist. Perhaps I am drawn to Boas because, like him, my trajectory seems to be somewhat erratic: I began in psychology, incorporated anthropology, grew even further through psychology, and am now taking both fields into any area of study about which I find myself passionate. Like Boas, my work has centered around the experience of internal processes of meaning-making in relation to the world around us. As I acquired a better understanding of psychology and science in general, I have realized a desire to apply the approaches I've learned and look through lenses of both in a more reflexive, self-aware way. Even now, as I contribute what is the first publication of my early career, I have strongly felt what I imagine Boas was experiencing throughout his own process.

Feminist Contributions to a Psychology of Science

While this essay is not focused on women in science, as a woman in science providing an analysis of the situatedness which contributes to all research, I would consider it negligent if I did not consider my own positionality as a researcher on this topic. A significant aspect of such a position is that I consider myself to be a feminist scholar. This is an aspect of my academic experience that I see to be closely aligned with the work presented here. Therefore, it is important for me that I bring to attention the absence of women's influence throughout much of what we consider to be science due to the long history of gatekeeping that prevented women (as well as people of color) from contributing to academia. Present within the same power structure, which operates within the binaries of male/female, are also the dichotomies of rational/empirical, natural/social sciences, and objective/subjective. I would hope that if Boas had been aware of the advancements in feminist scholarship, he might find himself working to step outside of such a structure in order to deconstruct these binaries even more than he was already doing during his time, but such assumptions cannot be made.

Boas did not directly address the issue of women in science, but he certainly had an influence on women's role in anthropology. And I am grateful to have been taught about his influence by a woman in the field. Furthermore, as a woman in the social sciences, writing at an intersection between psychology and anthropology, I write with interdisciplinary feminist scholarship in mind as it has relevance to the present work. This commitment foregrounds the intentional reflexivity and subjectivity with which I approach the study of science. Harding speaks of the criticisms that feminist scholars have faced in their contributions to science, especially that:

> women are inferior at rigorous observation and at reason, such critics are probably thinking, and-worse yet-feminism is a politics. How could women and politics be producing facts that anyone should regard as serious challenges to the impersonal, objective, dispassionate, value-free facts that the natural and social sciences have produced?
>
> *Harding, 1991, p. 108*

Harding explains that there is a paradoxical nature to the way in which we view feminist and other reflexive perspectives in science as harmful to the objective nature of the discipline. In actuality, it makes the discipline stronger by directly addressing the ways in which we influence our work. Yet reflexivity and other valued qualities of the social sciences are still viewed through a sexist, and therefore biased, lens:

> When we dub the objective sciences "hard" as opposed to the softer, i.e., more subjective, branches of knowledge, we implicitly invoke a sexual metaphor, in which "hard" is of course masculine and "soft," feminine. Quite generally, facts are "hard," feelings "soft."
>
> *Keller, 1983, p. 188*

Keller further elaborates on this belief as she writes, "the virtual silence of at least the nonfeminist academic community on this subject suggests that the association of masculinity with scientific thought has the status of a myth which either cannot or should not be examined seriously" (Keller, 1983, p. 187). Keller continues, "Unexamined myths, wherever they survive, have a subterranean potency;

they affect our thinking in ways we are not aware of, and to the extent that we lack awareness, our capacity to resist their influence is undermined" (Keller, 1983, p. 187). Harding suggests feminist empiricism, developed within biology and fields of the social sciences, as one possible strategy for combatting these types of assumptions as well as the biases and prejudice that accompany them within research of which we must become more aware to produce better, more rigorous scholarship.

Although his time in the social sciences took place long before Harding's writing on the topic, Boas was a scholar who I believe engaged in feminist empiricism before its true emergence. While he could not have understood what such an approach would become in other fields not yet developed, Boas confronted sexism through his teaching as well as scientific racism through more rigorous, data-driven research, utilizing oft-misused methodologies in order to make a valuable statement against the intellectual zeitgeist at the time. However, it is the intent of his work that gives it the quality which I wish to retroactively apply to his career as feminist. It is important to understand that "feminist empiricism appears to leave intact much of scientists' and philosophers' conventional understanding of the principles of adequate scientific research. It appears to challenge mainly the incomplete practice of the scientific method, not the norms of science themselves" (Harding, 1991, p. 113). Therefore, Boas' work was not necessarily different from the scholarship of other individuals who we look to as doing good science, and it is difficult to identify whether he was aware of the antiracist qualities of his work. But what distinguishes his scientific contribution is its juxtaposition to prevailing discussions of race in academia, and thus its impact on the lives of individuals who were being harmed by faulty science.

Addressing our situatedness makes us responsible for our work as well as the effects it has. I wish to close this section with one last contribution from Haraway as she warns against the pitfall of believing that science can be entirely objective and without human contamination, "only the god trick is forbidden" (Haraway, 1988, p. 589). She explains, "feminism loves another science: the sciences and politics of interpretation, translation, stuttering, and the partially understood" (Haraway, 1988, p. 589). I hope it will be evident to the reader the present work is also one that is but one interpretation and could have been taken up in any number of differing ways by other scholars, and therefore it may be seen as flawed by many, but they must then take stock of their own interpretation to inform their own work. It is my hope in presenting such a contribution that we will continue to question naïve views of the objectivity of science and what parts of ourselves we bring to our work.

Final Thoughts

This analysis is by no means a complete view of Franz Boas, his work, the theories discussed throughout, or the role of racism nor antiracism within science. However, through even this imperfect and incomplete work, it is clear that Boas had an incredible impact on the way in which we view and study race today.

However, these contributions must be considered in conjunction with the personal, cognitive, and social processes that influenced Boas' work and role in science. Psychology of science provides a unique, holistic perspective for considering the work of scientists, not merely as idealized thinkers, but as human beings influenced by the experiences in life and the world.

QUESTIONS FOR DISCUSSION

I include as questions for discussion a set Harding asks to probe taken for granted assumptions about the nature of science and objectivity. There are relevant to a psychological exploration of the work of Franz Boas and his contributions to anthropology and to science more generally:

1. What kinds of things can be known? Can "historical truths," socially situated truths, count as knowledge? Should all such situated knowledges be regarded as equally plausible or valid?
2. What is the nature of objectivity? Does it require "point-of-viewlessness"? How can we distinguish between how we want the world to be and how it is if objectivity does not require value-neutrality?
3. What is the appropriate relationship between the researcher and her or his research subjects? Must the researcher be disinterested, dispassionate, and socially invisible to the subject?
4. What should be the purposes of the pursuit of knowledge?
5. Can there be "disinterested knowledge" in a society that is deeply stratified by gender, race, and class? (Harding, 1991, p. 109)

References

Baker, L. D. (1998–1999). Columbia University's Franz Boas: He led the undoing of scientific racism. *The Journal of Blacks in Higher Education, 22*(1), pp. 89–96.

Boas, F. (1969/1928). *Materials for the study of inheritance in man.* New York, NY: AMS Press.

Boas, F. (2011/1911). *Changes in the bodily form of descendants of immigrants.* Redditch: Read Books LTD.

Carré, D. (2019). Toward a cultural psychology of science. *Culture & Psychology, 25*(1), pp. 3–32.

Darnell, R. (2016). Who was Franz Boas? How do we know? And why should we care? *General Anthropology, 24*(1), pp. 1–7.

Grosul, M. & Feist, G. J. (2014). The creative person in science. *Psychology of Aesthetics, Creativity, and the Arts, 8*(1), pp. 30–43.

Haig, B. D. (2005). An abductive theory of scientific method. *Psychological Methods, 10*(4), pp. 371–388.

Haraway, D. J. (1988). Situated knowledges: The science question in feminism and the privilege of partial perspective. *Feminist Studies, 14*(3), pp. 575–599.

Harding, S. (1991). What is feminist epistemology? In S. Harding (Ed.), *Whose science? Whose knowledge? Thinking from women's lives*, pp. 105–137. Ithaca, NY: Cornell University Press.

Keller, E. F. (1983). Gender and science. In S. Harding & M. Hintikka (Eds.), *Feminist perspectives on epistemology, metaphysics, methodology, and philosophy of science*, pp. 187–286. Dordrecht: Reidel Publishing Co.

King, C. (2019). *Gods of the upper air: How a circle of renegade anthropologists reinvented race, sex, and gender in the twentieth century.* New York, NY: Anchor Books.

Lewis, H. S. (2001). The passion of Franz Boas. *American Anthropologist, 103*(2), pp. 447–467.

Singer, B. F. (1971). Toward a psychology of science. *American Psychologist, 26*(11), pp. 1010–1015.

7

RACHEL CARSON

A Scientist of Life Itself

Peder Schillemat

Personal Preamble

I recently encountered the argument, now several decades old, that the work of a psychologist is, in a sense, irresponsible when investigating small questions instead of concerning itself with the issues of human survival (Bevan, 1982). It reminded me of why I was so drawn to write this chapter on Rachel Carson. I first heard about Carson from a colleague much too late in my life. What my colleague explained about her life thoroughly intrigued me. She seemed to me the type of person who was not interested in just getting by in life. Instead, she really wanted to make an impact. As I got the chance to research her life further, I realized that with that drive came many different challenges impeding her ability to finally publish what really mattered to her. My interest in Rachel Carson is rooted in my admiration for the work that she achieved. Carson accomplished all she did while dealing with the challenges of being a woman in science and contending with people and institutions who did not want her work to be made public, and in fact actively tried to stop her. She recognized the deep problems in the society and time that she was living in and sought to reform them. This, I believe, is one of the extraordinary opportunities for scientists and one of the reasons that I chose to write about the psychological aspects of Carson's life. In order to have even a small hint of the positive impact on the environment that Carson has had, I feel I will need to understand the various factors that create the opportunity for a scientist to change the world for the better.

Acknowledgment of Source Material

Throughout this paper, I am indebted to the scholarship that has previously been done to describe Carson's life. In particular, Linda Lear and her extensive reporting

DOI: 10.4324/9781003276692-9

and compilation of Carson's work help to ground the narrative throughout much of this work so that I can provide insight into the psychological aspects of Carson's work (Lear, 1998). Her work is heavily cited to provide the context necessary to produce this chapter.

Introduction

Rachel Carson is an important case study for reflecting on what it means to be a scientist. While she did not have an academic position or even a Doctoral degree, her work is unequivocally scientific and was influential in altering public policy and the perception of how humans interact with the natural world. In this chapter, I will be detailing Carson's life with a specific focus on how she presents an important case for the psychology of science and how science is practiced. The chapter will cover the different facets of Carson's life: personal, including her familial relationships, financial strains, and personal beliefs; sociocultural, including the impact of world wars, the role of women in science, as well as the political climate surrounding her work; and finally, the cognitive aspects of her work, especially her use of analogical and inductive reasoning to form her hypotheses.

This chapter will discuss all of these dimensions by following Carson's life sequentially. This is to offer a sense of who Carson is as a person and scientist and to situate the personal, cultural, and psychological aspects of her life and work. Section headings are used to illustrate the most relevant themes throughout Carson's biography and how they are implicated in her scientific work.

The Personal Dimension of Rachel Carson's Work as a Scientist

Rachel Carson stated that the "materials of science are the materials of life itself" (Carson, 1952 as cited in Lear, 1998, p. 218). This came after Carson received the John Burroughs Medal for her second work *The Sea Around Us* in 1951. In Personal Knowledge, Michael Polanyi (2015/1958) wrote, "The freedom of the subjective person to do as he pleases is overruled by the freedom of the responsible person to act as he must" (p. 309). The concern with acting responsibly can be seen throughout Carson's life as she struggled with the effort to be as involved in her research as possible amid recurring financial, familial, and academic constraints. These struggles are important to understand Carson as a person and scientist. Moreover, they illustrate the important insight that science doesn't take place in a vacuum: it is, in an important way, a deeply personal commitment and its practice is affected by myriad forces.

Carson openly acknowledged the constraints that a scientist must contend with, but also the way a personal belief can embolden a research topic. She wrote to her friend Dorothy Freeman while developing one of her later books, "I have no more energy than anyone who loves what he is doing...." (Carson, 1959, quoted in Carson & Freeman, 2022, p. 287) Important to understand is that Carson was battling cancer during this time. Moreover, she was already getting pushback on her research. Yet, a scientist's personal commitment can help to counter and resist the social and cultural constraints. Carson's love for her work, and her connection of this love with the energy that kept her going in it relates to Polanyi's emphasis on the scientist's personal passion closely connected to the dedication that makes scientific achievement possible: "[t]he personal participation of the knower in the knowledge he believes himself to possess takes place within a flow of passion. We recognize intellectual beauty as a guide to discovery and as a mark of truth" (Polanyi, 2015/1998, p. 300).

In the following pages, I will be describing the personal aspects of Carson's life that both allowed her to practice science and some of the ways her beliefs, her family, and financial constraints that all affected her work.

Rachel Carson was born in Springdale, Pennsylvania, to Maria and Robert Carson. She was the third child born much later than her other two siblings. She was not very close with her siblings but spent much of her childhood with her mother. The interactions with her mother were often outdoors on their land, where they would spend time paying attention to the birds and other wildlife. Carson's childhood instilled a persistent belief that one could come closer to God through nature, in part through the explorations of the natural world in which she engaged with her mother (Lear, 1998).

Carson was born to older parents who often struggled to find work to support the family. Robert Carson worked in the local mills but could not physically continue to work when old age hit. Although never destitute, the family's financial struggles did limit Rachel's educational options. For example, she had to wait to go to her local high school because of the cost of attending. This, nonetheless, did not deter her, as she continued to excel in her academic work within the conditions in which she found herself (Lear, 1998).

Carson's ability to excel was most evident in her writing. As a child, she had a skill in writing stories and gained confidence as she was encouraged to share and send her work to magazines. Her first-ever paid work was in a children's magazine where she wrote a war story (Lear, 1998). She described that time as one of the first moments of real confidence in her work. Because of this, she continued to develop her writing, which eventually led her to her undergraduate degree in English at the Pennsylvania College for Women.

Carson's time at college was interesting as she began to take time away from her family. For much of her early life, she was attached very closely to her mother, so it was a chance for her to be gone throughout the week from her parents. She would either go home on the weekends to see her mother, or her mother would come to see her at college, so they kept the close bond. This separation was where Carson started to develop some friendships, ones that her mother monitored very closely. While Carson did gain more independence, she continued to feel her mother's controlling interest in those with whom Carson associated throughout much of her life (Lear, 1998).

When Carson returned to spend summers with her family, she was responsible for a full household. Her father was old and could not work, and her older sister had had a child that Carson was often asked to care for. The time back home was difficult, and Carson could not wait to get back to her studies. Her parents sold land plots to the University as capital in order to enable Rachel to continue to attend classes. Despite the additional strain this placed on the family, the arrangement allowed Carson the freedom to study as she liked (Lear, 1998).

Carson excelled in her writing classes especially and often presented very poignant works that began to hint at her involvement with the natural world. But it was not until she later took a course with Mary Scott Skinker that she felt the urge to shift toward a scientific career. Skinker was described as a bold woman, rigorous in what she expected from her students because she, herself, needed to uphold that level of rigor for her own work. She never married and made it clear that she had no intention of marrying as she wanted to continue her research. This was uncommon at the time but not unheard of. It did make it difficult for Skinker at the Women's College since there were differing ideals concerning the purpose of the education. While Skinker advocated that education should prepare women to gain a skill leading to a career, some of the administrators felt that an education was important for women only in so far as it contributed to establishing and maintaining a household (Lear, 1998). Skinker made it clear that she had no intention of letting her class be easy or valued as a quick passage into biology. She expected her students to learn and make contributions to biology. This was quite the change for Carson and one that she readily embraced. Not soon after taking Skinker's biology course, Carson switched her major from writing to biology. As she did so, she and Skinker became close friends, and Skinker helped to advance Carson's career forward.

Some of Carson's first scientific work available to her through this relationship. She was invited to study at the Marine Biology Lab in Woods Hole, Massachusetts. It was here that Carson began to increase her love of the biological world and her love of the sea. Carson spent hours studying and getting the chance to meet many of her future scientific colleagues. It was an incredibly fruitful time for her, one that she looked back on fondly (Lear, 1998).

After graduating from the Pennsylvania College for Women, Carson decided to continue her education with a Master's in Zoology at John Hopkins, a move that meant that she would be completely separated from her family. For Carson, it was both enabling and difficult. Because she had such a close connection with her mother, she did miss her, but the move also allowed her more freedom to connect with others her age without her mother's critical judgment (Lear, 1998).

Carson did exemplary work in her classes but found it difficult to carry out her research. As she was scrambling to narrow down her topic and draw conclusions from her own research, she also struggled financially. Her family could only support her so much, so Carson took on other jobs. She started teaching jointly with another colleague, which helped pay for some of her tuition, but things remained very tight. When Carson eventually approached the end of her master's degree, she hoped to continue toward a doctorate, but her financial strain prevented this. Carson's financial difficulties could have ended her scientific career before it began: this is a personal dimension to the scientific work itself, one that is incredibly important in her case. Many who do not have the finances to pursue advanced degrees do leave science behind for a more stable job and income.

Carson was fortunate in that her personal connection with Skinker helped to keep her engaged in science and in pursuit of a scientific career (Lear, 1998). On Carson's behalf, Skinker contacted the head of the Department of Fisheries and Wildlife, Elmer Wiggins, to get Carson a position as a field scientist (Lear, 1998). In this position, Carson would be able to not only work and have the financial support that she needed, but she would also be able to continue working in the field of science, albeit in a government role. At this time, Carson's father had passed away and Carson was solely responsible for supporting herself and her mother. The work did not provide a worry-free lifestyle but it afforded some security for Carson and her family.

Her role in field science was perfectly suited to Carson's skills in writing and her affinity for nature. She was able to visit rivers and write informational pieces that were heralded for the way she made the natural world come alive. Her most challenging issues with the position were often the bureaucratic duties assigned to her that would keep her away from writing and being out in the natural world she loved. This was no more significant than dealing with colleagues that would harp over the cost of paper and effectively shut down operations for her work. She knew that she had valued many of the opportunities that she had there, but she felt stifled in a position that she knew did not afford her the options to write and research freely (Lear, 1998).

Carson first tried to reach out to various publications in the hope of publishing in magazines to free time from her work in the Department. Her first submitted pieces were not accepted, so she had to move onto new avenues to find a way to continue her research. She decided to begin conducting research for her first book, *Under the Sea Wind* in 1941. She knew at the time that she would need an agent to help her get her book published. She hired Marie Rodell, who became an excellent agent for Carson and a colleague and friend who could advocate for her intensely. Rodell secured a book deal with Simon and Schuster for Carson's first book (Lear, 1998). It was a fascinating time for her, and Carson worked with fervor. As Carson moved farther and farther toward writing, her career in science was being affected by various societal and cultural factors that are important to note.

Sociocultural Aspects Surrounding Carson's Work

At this point, Carson was becoming more and more affected by the social and political world around her. The work of a scientist is always embedded in a cultural, political, and historical context. But perhaps Carson was especially impacted, as she wrote from the very consequential period of the 1940s to the 1960s. The timing of wars, resistance to even the discussion of the impact of insecticide use, and her perceived right to practice science as a woman were a part of the context and the dialogue. Moreover, she continued to experience personal difficulties.

When Carson was able to release *Under the Sea Wind* in 1941, it was a flop, mainly attributed to the timing of the US involvement in WW II (Lear, 1998). The US was entering the war, and her book was not as openly publicized. She received little recognition for the work and had to continue to worry about finances and whether her office would be shut down due to the war. There was also a concern that once the war ended, a man would be hired to replace her in her position as a field scientist. Amid these concerns, Carson still felt a desire to write and produce works that could speak to a deeper understanding of the natural world.

Her second book, *The Sea Around Us* in 1951, expanded on her earlier work, but it focused more specifically on the oceans. She continued to work with Marie Rodell, who helped Carson to secure a book deal once again. The book this time was published at an incredibly fortuitous time. It was 1951, a few years after WW II had ended and the Korean war had begun. Many Americans had been reeling from the wars, as well as the rise of nuclear weaponry and McCarthyism. The release of Carson's book came as an escape from these realities and allowed many Americans to revel in her poetic writing but also the detail she ascribed to the natural world. Because she never viewed her writing as an escape from reality but rather as a way of speaking more directly to it, Carson undoubtedly disagreed with this sentiment (Lear, 1998). In any case, the timing of the release was pivotal, and *The Sea Around Us* was read widely across the country.

Because of the success of *The Sea Around Us*, Carson finally had some financial security. She no longer had to work at the Department of Fisheries and Wildlife and ended up leaving her position. She had loved it, but the overwhelming bureaucratic responsibilities often kept her from doing more of the meaningful work that she did while out in the field or when writing. This new financial security gave Carson the chance to do both while also letting her agent deal with more of the financial arrangements and scheduling related to her writing.

Carson enjoyed the positive reviews and accolades that *The Sea Around Us* provided. Some of the early reviews spoke to the beautiful way that Carson was able to write about scientific principles that did not anthropomorphize nature but spoke to its beauty. Some of the reviews at the time said that it was "a brilliant study of the sea ... not only a superb example of scientific reporting but also a work of art" (Martin, 1951 as cited in Lear, 1998, p. 206). Another said that it was "a first-rate scientific tract with the charm of an elegant novelist and the lyric persuasiveness of a poet" (Breit, 1951 as cited in Lear, 1998, p. 205).

The favorable response to Carson's work focused on both her scientific acumen and her effectiveness as a writer. However, there was also criticism, much of which focused on her position and responsibility as a scientist. While these opinions were not as frequently expressed, some scientists felt that her book did too much to illustrate the beauty in the natural world – that it was too aesthetically expressive for a scientific work. Carson, though defended herself on this front and wrote, "the sea, and the moving tides are what they are. If there is wonder and beauty and majesty in them, science will discover these qualities" (Carson, 1952 as cited in Lear, 1998, p. 219).

Carson as a Woman in Science

What was most infuriating at the time, though, was the focus on her gender. One reader thought that the author must be a man due to the command of the material and scientific knowledge (Carson, 1954 as cited in Lear, 1998). For those who did not think that Carson was a man, there was an impression that if she was a woman, she must be a hulky one at that, and one who exuded the sterility often seen to be in keeping with the work of science. This was a view that seemed to be shared even by women. As a parody, Carson's friend Shelley Briggs drew a cartoon at the time an image that she felt embodied what Carson's audience thought she would be like. It "depicted her as a female of Amazonian proportions striding the seas, long hair tossing in the wind, and octopus in one hand, sea spear in the other" (Briggs, 1951 as cited in Lear, 1998, p. 207). Carson's sex also came up in interviews and TV appearances where the interviewers and audiences were often taken aback by her small size and femininity. The way in which Carson was perceived and described echoes Keller's point that science itself is often gendered, specifically seen as a male pursuit, and thus women who do practice science have been historically viewed as unfeminine or even nonsexual beings (Keller & Scharff-Goldhaber, 1987).

As difficult as it was to have her scientific work reduced to her gender, Carson was able to take moments during her rise to fame to discuss what she really felt about her commitment to science and literature. She stated,

> the aim of science is to discover and illuminate truth. And that I take it is the aim of literature, whether biography or history or fiction. It seems to me, then, that there can be no separate literature of science.
>
> *Carson, 1952 as cited in Lear, 1998, p. 219*

With this strong statement, Carson made it clear of her role as a scientist and an author and that these roles were not mutually exclusive but instead worked in tandem with one another. She would describe in other responses that "if there is any poetry in my book about the sea, it is not because I deliberately put it there, but because no one could write truthfully about the sea and leave out the poetry" (Carson, 1952 as cited in Lear, 1998, p. 219).

While Carson would have liked to revel more in the success of her work, she was still beset by personal commitments to her family. Her niece, Marjorie, had become pregnant through an affair with a married man. This was something that Carson and her family needed to keep from the public eye, especially as Carson was rising in fame. Carson's familial obligations often had a considerable influence on how much Carson could do (Lear, 1998). She knew that as her fame started to arise, she would also have to make sure that there was no scandal surrounding her and taking the focus off the message of her work.

Political and Environmental Issues Surrounding Carson's Work

While Carson was dedicated to making connections between the mysteries and wonders of the natural world, she was also very conscious of the forces threatening to destroy it. After writing *The Sea Around Us*, Carson bought a home in Maine to be out in the forests and close to the sea. But she knew that this land was under threat of development. She strove to keep it available in its natural state, not only for herself but for others to enjoy. She worked with her friend Dorothy Freeman to speak with owners around the forest and enlist their participation in preserving the land. As such, Carson actively worked toward preservation issues at her residence in Maine (Lear, 1998).

But nothing was a clearer message to Carson that the natural world was under siege than the spread of pesticides. DDT (Dichlorodiphenyltrichloroethane) was heralded at the time as being a miracle of science. It would make everything pleasant for those out in nature, enabling them to enjoy it without the intrusion of annoying mosquitoes or bugs (Lear, 1998). But Carson was increasingly aware of other effects. She wrote in the beginning of *Silent Spring* that the proliferation of chemical pesticides would not lead to the hoped-for utopia but instead to the silencing of birds, the loss of natural landscapes, and concerning health issues for many (Carson, 2002).

Carson was aware of the political implications of writing such a book. She knew that many people would view her as being against agriculture and the progress of technology. Carson had her reservations as she wrote:

> I shut my mind, … refused to acknowledge what I could help seeing. But that does no good, and I have now opened my eyes and my mind. I may not like what I see but it does no good to ignore it, and it's worse to go on repeating the old "eternal versities" that are no more eternal than the hills of the poets.
>
> *Carson, 1958a, as cited in Carson & Freeman, 2022, p. 249*

Carson, however, was clear that she needed to speak to the issues present, and that checks and regulations were imperative. Without them, we were on a path that led to disaster.

As expected, there were critics. Many people viewed her research into pesticides as unfounded. Arguments were made that there was no proof that Carson could deductively prove that pesticides were causing problems for the natural world. She was encouraged to simply leave the issue alone. This was not an accommodation that Carson could make. She was not willing to sit back when she had such convictions about the effects of which she was aware through her analysis of future directions and probabilities. Instead, she concluded that others in the scientific community were resisting her conclusions for reasons that were not entirely objective:

> I'm convinced there is a psychological angle in all this: that people, especially men, are uncomfortable coming out against something, especially if they haven't absolute proof that something is wrong, but only a good suspicion. So, they will go along with a program about which they privately have acute misgivings. So, I think it is most important to build up the positive alternatives.
>
> *Carson, 1958b, as cited in Lear, 1998, p. 335*

With her resolve moving her forward, Carson directed her efforts to investigating and communicating the effects that pesticides had on the environment and the animals that lived there, as well as the detrimental effects on humans. During this time, a drug was given to many pregnant women to help with nausea. But the side effects of this drug were not fully known until many babies were born with physical deformities, including missing limbs. Perhaps this made people more open to Carson's arguments. She wanted to make it clear that the emphasis we place on chemically treating our bodies can have consequences, and that similar risks apply to our chemical treatment of the natural world. She wanted to emphasize that all people should be concerned about the human impact on nature, not just those that felt a certain fondness for it and liked to spend time outside (Lear, 1998).

Continuing Personal Difficulties Impacting Carson's Work

Carson was moving as quickly as she could to draft *Silent Spring*. This was incredibly difficult as Carson had many personal issues to confront throughout the writing process. Before beginning the book, Carson's niece Marjorie had passed away. This created a new difficulty as Marjorie's son Roger did not have any other family to turn to and was then left in the care of Carson and her mother (Lear, 1998). Roger was said to be an impulsive and challenging child who often had difficulty staying still. Carson would do as much as she could with him, trying to get out and explore the sea and the forests, being as available as possible, but it took a toll on what she could do in developing her book. She would often try and get him out of the house or have her mother take care of Roger, but this was also incredibly difficult as Marie Carson was also becoming much older and in need of full-time care (Lear, 1998). As her health declined, it was important that Rachel find an appropriate nurse that her mother liked and respected. Carson was eventually able to find an adequate caretaker, which should have freed up more time to write, but unfortunately, she faced new difficulties.

Carson herself had health issues that were genetic but exacerbated by the medical practices of her time. She was diagnosed with breast cancer, with which her mother had also been diagnosed. Astoundingly, Carson did not hear about her diagnosis until it was nearly impossible to treat. The first doctor who noticed the cancer did not inform her of it. At that time, it was customary for any poor diagnoses to be shared only with the man of the household, who would be tasked with sharing the information with his wife. As Carson never married, her doctor did not tell her that the growth was malignant, and as such, she received no treatment and it continued to worsen (Lear, 1998).

Her health issues, in turn, kept Carson from being able to go out and do as much fieldwork as she had been able to do in the past. She was more homebound, and some of the treatments that she had to endure when dealing with her increasing illness incapacitated her ability even to read (Lear, 1998). She relied very much on her research assistant and her friends to support her in moving forward.

Her relationships, however, also added their own complications. Her friend Dorothy Freeman was the most intimate relationship of Carson's entire life. They were so intimate that they felt it imperative to make sure that Freeman's husband was comfortable with their closeness. Dorothy and Carson encouraged each other and had confidence in the work each was pursuing. However, the development of *Silent Spring* though was difficult for Freeman to understand because it did not relate to the wonder and beauty that she and Carson both felt in nature. Instead, it was designed to highlight the ugliness that could be found in humans' pursuit to control nature. It took Freeman time to understand why Carson felt such an obligation to write *Silent Spring* (Lear, 1998). Carson struggled for a while without the support of her friend (who eventually came to appreciate the book).

Despite all these personal aspects that made it difficult for Carson to write *Silent Spring*, she developed the book and got it published by June 30, 1962. It was widely received, and the sentiment at the time was that the publishing of *Silent Spring* led to a noisy summer, full of debate and discussion (Lear, 1998). This noisiness was no more apparent than for Carson, who was called on by the media, government, and many others about her work.

Political Impact of *Silent Spring*

Some of the loudest voices were of those going to feel the most economic loss from the public taking seriously the hypothesis that Carson laid out in the book. Many in the pesticide industry sought to discredit Carson as nothing more than an emotional female alarmist. Interesting to note is that often they would attack her character and who she was as a person rather than the science she presented in the book. The head of agriculture at the time stated, Why [would] a spinster with no children [be] so concerned about genetics? (Benson, n.d. as cited in Lear, 1998, p. 429). It was also added that she was probably a Communist.

The criticisms that Carson received are interesting to highlight, especially considering how she was perceived as a scientist when she published *The Sea Around Us*. In that book's publication, she was confronted by the opinion that because she was a woman, and a feminine and a small one at that, she could not have written such beautiful and concise scientific prose. The critiques she received when she published *Silent Spring* stated something similar but now with a tone more denigrating toward her status as a woman, focusing on the fact that she was not a mother, or that she was overly emotional, just taken up by the alarm around her (Lear, 1998). Important to note is that Carson was ready to combat these claims. Since there were no criticisms of her science itself, she stood firm. During this time, she also did her best to avoid looking frail despite her growing health issues. She wanted to make evident that she was both a woman and a scientist.

Eventually, Carson was called on by Congress to discuss more about the effects of chemicals on the environment. While being reserved and bold, she made clear the effects that pesticides could have on the environment and human life. The claims that she made were not a form of alarmism, nor were they only the opinions of someone interested in nature but instead the reasoned conclusions of a person who had spent her life being a part of nature and observing its connection with humanity. After her testimony to Congress, government research was conducted on the effects of pesticides and the environment and found all of Carson's claims valid. This led to further protection for the environment and the creation of new protection agencies such as the EPA (Lear, 1998).

Scientific Reasoning in Carson's Work (Cognitive Aspects)

All that Carson was able to do would not have been possible if not for the scientific reasoning that was part of her work. Throughout her work, Carson illustrated the use of inductive and analogical reasoning to bring to the fore the realities and mysteries that were present in the world around her, and which she wanted to communicate to others. Moreover, Carson did not focus solely on proving something particular but instead questioned ways that science was being practiced and applied. She did not rely solely on facts and numbers but used analogies to extend the scope of her work and make important concepts deeper and experientially vivid.

Inductive Reasoning

One of the first things to understand is Carson's use of inductive reasoning. Inductive reasoning is based on moving from the particular to the general, drawing on multiple sources of data (multiple instances) to then make a claim that has a high probability of being true (Giere, 1991). It also has a possibility of being false, which speaks more to the need to continue to question and examine data and keep searching for the highest probabilities. Carson's use of inductive reasoning is evident throughout her fieldwork. When studying marine life, she would investigate "positive alternatives" to try to understand the mysteries that were there (Carson, 1958b as cited in Lear, 1998, p. 335). She was vigilant in her representation of the facts of the natural world. She disliked works that would anthropomorphize nature to make it more understandable. She instead relied on making sure the claims she made in relation to any scientific phenomenon were grounded in solid research.

One illustration of her inductive reasoning is evident in her publication of *Silent Spring*. Carson was dedicated throughout the work of *Silent Spring* to investigating the effects of pesticides as thoroughly as possible. To carry out this level of research, she continued to ask questions and consider possible consequences to the effects of pesticide use. The first chapter of *Silent Spring* even highlights a potential effect of the use of all these pesticides by offering a harrowing and somber look into a fictional town where the land is desolate and birds do not sing. This hypothetical town serves as a kind of thought experiment, allowing her to make inferences about the consequences of continued pesticide use. She then goes about asking the important questions around pesticide use and presents research into how it has affected humans, how it has affected the environment, and how she sees that it could continue to alter the landscape. This is not proof, but it serves a hugely important role in science. It is a way of continuing to ask questions and encouraging others to ask them as well. New questions then enable the creation of tests and experiments to move into more deductive reasoning. Because of Carson's position, she often could not perform all the investigative work herself, but her invitation to others to ask questions was an important step in making the investigation possible.

Analogical Reasoning

In conjunction with her ability to ask questions and pry further into the phenomena, Carson created analogies to help build up the wonder and despair that she found around her. The use of analogies in scientific reasoning has been theorized as having two major principal functions. First, "that analogies justify pursuit by supporting plausibility arguments" and second, "that analogies can serve as a guide to potential theoretical unification" (Nyrup, 2020, p. 1). Carson's use of analogies was both a form of illustrating the wonder of the world and supporting an argument about the interconnectedness of nature and humanity. For example, Carson challenged parents with an analogy comparing disregard for nature's wonders to physical blindness. She stated, "For most of us, knowledge of our world comes largely through sight, yet we look about with such unseeing eyes that we are partially blind. One way to open your eyes to unnoticed beauty is to ask yourself, 'What if I had never seen this before? What if I knew I would never see it again?'" (Carson, 1988, p. 52).

As Carson wrote about the wonder of the natural world, she tied in these questions that asked people to consider further their connection with the beauty of the world and used analogies from fantasy realms to make the natural world more accessible to her readers. In her writing, she used terms such as "fairies" (p. 45), "that the lichens have the quality of fairyland" (p. 36), "this one must be a tree for elves" (p. 40). It is important to note, though that Carson did not anthropomorphize the natural world. This is perhaps clearest in her descriptions of the sea, where she does not attribute personal attributes to the sea life she studies but tries to chronicle it exactly. The analogies and metaphors she used were intended to emphasize further the need to continue investigating the mysteries of nature, not to provide an explanation. Nevertheless, scientific analogies allow both the scientist and her audience to think about the world in a different way, encouraging continued thinking and new insights.

Conclusion

A question central to this chapter and the book as a whole is what it means to be a scientist or to do science, which is not discussed often in psychology. But if psychology is a science, patterned after the natural sciences, this question is obviously relevant to everything we do. One might glibly state that to be a scientist is to practice science, which is obviously true. But "practicing science" takes many forms, and much of which is outside of the laboratory. Scientists are complex persons with a story, who feel deeply connected with their work and who always practice within a context that can influence the progress and impact of their work profoundly. Carson faced many obstacles that affected her ability to study science freely, including financial struggles, her personal commitments to her family, the cultural implications of practicing science as a woman, and the political resistance to exposing the harm caused by industry. Nevertheless, she did what she could, and it is important then that we consider that amid the many difficulties, many people do practice science in spite of the constraints, using the tools available to them, as Carson so powerfully demonstrated.

A second point to make about science inspired by this case study is that the communication of scientific knowledge, including in forms that are accessible to the public is an important aspect of doing science, as is providing information for creating policy. Carson utilized her skills in writing to make scientific work available publicly, and to convince government bodies of the important implications of her research. Her life and work challenge us to broaden the scope of how to think about and practice science. She summarizes it best:

> Many people have commented with surprise on the fact that a work of science should have a larger popular scale. But this notion that "science" is something that belongs in a separate compartment of its own, apart from everyday life, is one that I should like to challenge. We live in a scientific age;

yet we assume that knowledge of science is the prerogative of only a small number of human beings, isolated and priestlike in their laboratories. This is not true. It cannot be true. The materials of science are the materials of life itself. Science is part of the reality of living; it is the what, the how, and the why of everything in our experience. It is impossible to understanding man without understanding his environment and the forces that have molded him physically and mentally.

Carson, 1952 as cited in Lear, 1998, p. 218

QUESTIONS FOR DISCUSSION

1. Carson was deeply invested in her work as a scientist, both emotionally and as a moral cause. What questions does this raise about the place of values in science and scientific reasoning?
2. Carson was celebrated for the beauty of her descriptions of the natural world. What role should literary expression play in science? For example, does it enhance scientific description and facilitate understanding of scientific concepts or does it get in the way of explanation and clear communication?
3. Carson used the analogies of fairies and lichens throughout her work to help discuss the wonder of life around her. Do fantasy elements have any place in science? How did Carson use these analogies while still maintaining scientific rigor?
4. How are the financial and familial pressures that Carson experienced typical of scientific work? How are they atypical?

References

Benson, E. T. (n.d.) [Letter to Dwight D. Eisenhower].

Bevan, W. (1982). A sermon of sorts in three plus parts. *American Psychologist, 37*, 1303–1322.

Breit, H. (August 1951). Reader's choice. *The Atlantic Monthly, 188*, 82–84.

Briggs, S. (1951). Rachel as her readers seem to imagine her. [Illustration].

Carson, R. (1952, January 29). [Speech]. National Book Award.

Carson R. (1954, April 21). [Speech]. "The real world around us". Theta Sigma Phi.

Carson, R. (1958a, February 1). [Letter to Dorothy Freeman].

Carson, R. (1958b, September 24). [Letter to Paul Brooks].

Carson, R. (1959, October 18) [Letter to Dorothy Freeman] as written in Always, Rachel. (1994)

Carson, R. (2002). *Silent spring*. Boston, MA: Houghton Mifflin Harcourt.

Carson, R. (2011). *Under the sea wind*. New York: Open Road Media.

Carson, R., & Freeman, D. E. (2022). *Always, Rachel: The letters of Rachel Carson and Dorothy Freeman, 1952–1964*. New York: Open Road Media.

Carson, R., & Lee, K. (1998). *The sense of wonder.* New York: HarperCollins Publishers.

Giere, R. N. (1991). *Understanding scientific reasoning.* New York: Holt, Rinehart, & Winston.

Keller, E. F., & Scharff-Goldhaber, G. (1987). Reflections on gender and science. *American Journal of Physics, 55,* 284.

Lear, L. (1998). *Rachel Carson: Witness for nature.* New York: Macmillan.

Martin, E. H. (1951, June). Brilliant study of the sea. In *The Evening Sun.* Baltimore, MD: Tribune Publishing.

Nyrup, R. (2020). Of water drops and atomic nuclei: Analogies and pursuit worthiness in science. *The British Journal for the Philosophy of Science, 71*(3), 881–903.

Polanyi, M. (2015). *Personal knowledge: Towards a post-critical philosophy.* University of Chicago Press.

8

SCIENCE IN THE MISTS

Dian Fossey

Stephen L. Antczak

Introduction

My interest in Dian Fossey arose directly from the literature review for my dissertation proposal. In this proposal, I outlined a research project to learn about the
psychological effects that studying endangered species might have on conservation researchers. Since Dian Fossey famously studied one of the most endangered
species on the planet, mountain gorillas, it seemed obvious that I needed to at least
read her book, *Gorillas in the Mist*. Before reading it, I had only a vague notion
of who she was, and the research and conservation efforts she conducted. I knew
of, but had not seen, the movie based on her book. In fact, I don't watch fictional
accounts of noted scientists or mathematicians, as I find nonfictional accounts of
their work far more enticing.

In reading *Gorillas in the Mist*, however, I became enthralled. I was impressed
that someone with no training in conducting research, with no scientific training
whatsoever, really, could be thrown into the field with almost no preparation, and
somehow transform into a *bona fide* scientist—that is, one with the appropriate
academic credentials and career path—not to mention the world's foremost expert
on mountain gorillas at the time. I fell in love with Fossey's rough edges and
intense commitment to the gorillas she studied. I have always had a soft spot in
my heart for those who break the rules, and in doing so demonstrate that the rules
needed breaking, and that's what she did. She broke *all* the rules: the rules set by
the scientific establishment, the rules set by conservation groups, the rules set by
locals, and the rules of Academia. She pissed off, and still pisses off, a lot of people;
some of those people tried to have her arrested, deported, defunded, publicly
shamed, professionally embarrassed, and blamed for some of the terrible things
that happened to the mountain gorillas while she was there. Eventually, she was

DOI: 10.4324/9781003276692-10

murdered, probably for breaking at least one of those sets of rules. It is important to keep all this in mind when considering some of the controversies connected to her then, and those connected to her now, rather than cherry-picking specific abhorrent behaviors and damning her for them.

During my brief foray in the film industry, someone once told me, or perhaps I read somewhere, the following observation: If you're not being sued by someone, you're doing something wrong. Dian Fossey was definitely doing something right. Given the importance of Dian Fossey's work, and the knife's edge of survival upon which the mountain gorillas exist, the example of Dian Fossey, while far from exemplary, is, to me, worthy of respect both as a scientist and a conservationist.

It's difficult to imagine being given the opportunity to undertake, as principal investigator, a major field study of a population of critically endangered animals, located in a region in turmoil due to violent political upheaval, and being funded by an organization within the United States government, without having any research experience at all, and having only an undergraduate degree in a completely unrelated discipline. Yet, this is what Dian Fossey did, although it was not as cut-and-dry a progression. In some respects, Fossey was lucky, but she possessed the wherewithal to take advantage of that luck. Without the help of Louis Leakey, it never would have happened, but without her own determination to visit Africa the first time, she never would have met Leakey. Without her skills as a photographer and a writer, she would not have been able to show him the articles and accompanying photos that impressed him. She possessed intense dedication, bravery, and recklessness, as demonstrated by her having her appendix removed after being told by Leakey it was required, although it was later revealed as a test of her dedication, which Leakey had no intention of her actually carrying out. She might have been offered the job anyway, but certainly this helped make it an easier decision for Leakey (Dian Fossey Gorilla Fund International, n.d.; Fossey, 1983; Mowat, 1987).

It is in light of the appendix incident that Dian Fossey's work as a scientist and conservationist should be considered. In some ways, it portends her approach to research as well as to conservation: Dian Fossey was willing to do whatever she believed was necessary to study, and protect, the gorillas that became both her research subjects and her closest friends, no matter the cost to herself. The relationship she fostered with gorillas, and protecting them from poachers and the encroachment of human civilization, is what allowed her to derive meaning from her work as a researcher. In this sense, meaning consisted of her "subjective reasons for action" (O'Doherty, Osbeck, Schraube, & Yen, 2019, p. 16), both in regard to her research and her active conservation methods.

Major Scientific Achievements

Fossey's contributions to science are perhaps less obvious than the contributions of other scientists. She did not discover a new species or a new celestial body; she did not identify any fundamental laws of nature; she did not develop a new technology; she did not develop any major theories or research techniques. She did conduct the most accurate census of mountain gorillas in the Virungas that had yet been done, although much of this work was carried out by graduate students (Fossey, 1983; Mowat, 1987). She was instrumental in promoting a public image of the gorillas that was at odds with what it had been, that they were mostly gentle vegetarians as opposed to fierce hunters (Mowat, 1987). She perfected techniques for habituating gorillas with humans, which allowed her to observe natural gorilla behavior up close without being a major distraction to them (Fossey, 1983; Mowat, 1987).

Louis Leakey's interest in gorillas had been to see what observing their behaviors could teach us—"us" being humanity—about human behavior. He was a paleoanthropologist, after all (Leakey Foundation, n.d.). However, Fossey, it seemed, was more interested in studying gorillas for the gorillas' sake as opposed to any insights they might reveal about what it means to be human. Leakey was also responsible for recruiting Jane Goodall to the long-term research of chimpanzees, and Biruté Galdikas to study orangutans in Indonesia (Leakey Foundation, n.d.). Leakey apparently believed that women were more suited to such work, that women made better field biologists because they were more patient than men (Paulson, 2017), and their not having been trained as scientists would make them less constrained by the rigors of what was then considered proper research conduct (Goodall, 2010).

True or not, it is difficult to question their dedication to their work. Goodall has so far devoted over sixty years to the study and protection of chimpanzees (Jane Goodall Institute, n.d.); Galdikas has worked with orangutans since 1971 (Orangutan Foundation International, n.d.); and Fossey studied mountain gorillas for thirteen years (Fossey, 1983). As mentioned before, she was murdered, possibly due to her stance on protecting and studying gorillas versus her treatment of local humans who did not share her views on how best to do so (Mowat, 1987), which meant as few disturbances as possible to them and to their habitat (Fossey, 1983). Her views put her at odds, for a long time, with the idea of promoting tourism as a means of raising revenue for conservation efforts. She did not believe that gawking Europeans would contribute to the well-being of the gorillas, and would get in the way of legitimate research efforts (Fossey, 1983; Mowat, 1987).

While Fossey's activities included protecting the gorillas, she did manage to do research, as well. Her purely scientific contributions to the study of the mountain gorillas are best understood when looking at the titles of the academic papers published over the course of her research:

"Vocalizations of the mountain gorilla" (Fossey 1972), and "Observations on the home range of one group of mountain gorilla" (Fossey, 1974), appeared in the journal *Animal Behaviour*. "Feeding ecology of free-ranging mountain gorillas" (Fossey, 1977), appeared in a book called *Primate Ecology*. "Development of the mountain gorilla (*Gorilla gorilla beringei*): the first thirty-six months" (Fossey,

1979), appeared in a book called *The Great Apes*. "Reproduction among free-living gorillas" (Fossey, 1982) appeared in *The American Journal of Primatology*. "Infanticide in mountain gorillas (*Gorilla gorilla beringei*)" (Fossey, 1984), appeared in a book called *Infanticide: Comparative and Evolutionary Perspectives*. None of these titles promise groundbreaking findings to shake the foundations of science, biology, anthropology, evolutionary theory, or even primatology. They are each simply small pieces to the puzzle that is an understanding of gorilla behavior.

Fossey's major contribution was, arguably, in confirming and extending the methods of habituation that allowed her such close, intimate contact with "her" mountain gorillas, giving her access to observe their behaviors at close range. While this may seem to be counterintuitive, that her presence allowed her to observe natural gorillas behavior, this is exactly what habituation allowed. The gorillas settled into their normal behaviors once they perceived Fossey as non-threatening and the newness of her presence wore off.

Fossey's longest lasting contribution to the study of the mountain gorillas of the Virungas is probably due to her fundraising and "active conservation" (Fossey, 1983; Mowat, 1987) activities. Active conservation primarily centered on the removal of traps set by poachers to capture game animals for bushmeat, such as the small antelope-like duikur. The patrols of Fossey and her assistants resulted in the removal of thousands of such traps—usually consisting of snares—over the years. Gorillas could get caught in such traps, and even if they escaped, they could suffer wounds that could get infected and turn gangrenous, maiming them for life, and sometimes killing them. Other active conservation actions were directed against local farmers and their cattle, who would decimate the vegetation that gorillas fed on. Fossey was known to shoot cattle, killing some, or intentionally wounding them, making it too costly for the local farmers to allow their cattle to graze in the park (Fossey, 1983; Mowat, 1987).

Despite Fossey's antipoaching activities, several gorillas that she had been studying, and had grown close to, were killed, including the one she had become most fond of, Digit. His killing motivated her to begin the Digit Fund to raise money for increased active conservation efforts, but this proved problematic. As will be explicated later, her antipoaching activities were controversial, and her personality was such that anyone who did not agree with her views on the subject became an adversary. One wonders whether, had she been a man, she would not have been undermined in her attempts to increase active conservation efforts, but reports of her behavior were used to wrest control of the money donated to the Digit Fund away from her, and there followed a concerted effort to have her removed as director of the research facility at Karisoke (Mowat, 1987).

Aside from destroying traps, Fossey would also destroy anything belonging to poachers (Fossey, 1983; Mowat, 1987). When a poacher was captured by her hired patrols, the poacher would be bound, and then sometimes whipped with nettles, and Fossey would don a Halloween mask and curse the poacher before he was turned over to the authorities, so that among local peoples she acquired a reputation as a witch, among other things (Fossey, 1983; Mowat, 1987). This behavior, which cannot be condoned, is best understood in light of her dedication to protecting the gorillas and their habitat. Her active conservation tactics were highlighted by others in attempts to undermine her work as a scientist (Mowat, 1987), but this does beg the question: Can, or should, one's work as a conservation scientist be judged based solely on the veracity of the science itself? If the conservation activities cannot be separated from the research activities, it is a question worth asking.

Regarding those activities, according to the Dian Fossey Gorilla Fund International (n.d.) web site,

> We have shown, using our over 50-year database, that this type of daily presence in the forests is what is needed to protect these gorilla populations from the many threats they face, as well as to collect the information that is needed to provide the most effective conservation strategies. Tracker teams serve the role of both protection and data collection and are the key factor in saving the mountain gorilla population.

As recently as 2017, a three-year-old gorilla was freed from a snare. It seems that Fossey's strategy of daily patrols to destroy traps and run off poachers is still the best practice, even if her more questionable tactics aren't deployed anymore.

The most recent census, as of 2020, states that the population of mountain gorillas stands at 1063 (Dian Fossey Gorilla Fund International, n.d.). The 1981 census determined that there were only 242 free-living mountain gorillas (Fossey, 1983, p. 249). Considering that Fossey believed it was likely that mountain gorillas would be extinct in the wild by the end of the twentieth century, these numbers would indicate that her prescription of "active conservation" has worked. That there are any free-living mountain gorillas extant today may be considered Fossey's greatest accomplishment both as a scientist and as a conservationist.

Biographical Information

Fossey was born in San Francisco in 1932. She did not know her father, who had moved away early in her life. She was raised by a strict mother and stepfather. Her first attempt at college was to major in business, but after a summer stint working on a farm in Montana she decided she wanted to work with animals. She then changed colleges and majors to preveterinarian, but difficulty with physics and chemistry classes prevented her from continuing. She changed majors and schools again, finally getting a degree in occupational therapy in 1954. After working as an occupational therapist in California for a short time, she took the job as Director of Occupational Therapy at a children's hospital in Louisville, Kentucky, where she lived outside of town on a farm (Dian Fossey Gorilla Fund International, n.d.; Fossey, 1983; Mowat, 1987).

A friend's photographs and stories of a trip to Africa instilled in Fossey a desire to travel there as well. To pay for the trip, she used up her entire savings and took out a loan. During the trip she chanced to meet Louis Leakey, who suggested her to go and see the mountain gorillas of the Virungas. She did, and became enthralled.

She returned to Louisville and published in local papers several articles with photographs of her trip to Africa. Soon after, she met Leakey in Louisville where he gave a talk during a speaking tour of the United States and showed him the articles. Impressed, Leakey suggested to Fossey that she might be the right person to take up a long-term study of the mountain gorillas. He'd had great success with Jane Goodall, whom he'd gotten to take up a long-term study of chimpanzees six years earlier (Mowat, 1987; Goodall, 2010).

In 1966, Fossey was back in Africa to set up her first research station for studying mountain gorillas at Kabara in Zaire (later called the Democratic Republic of the Congo). The Virungas are mostly extinct volcanoes (two are active) that form a mountain range about 25 miles long and at most 12 miles wide, and make up the habitat of the mountain gorillas. They are mostly in the Democratic Republic of the Congo, although about 30,000 acres lie in Rwanda and a smaller portion lies in Uganda. All three countries have designated the area a national park or sanctuary for the gorillas (Fossey, 1983, p. xv).

Fossey spent only a short time at the Kabara facility. In 1967, political violence in the region resulted in her being forcibly removed from Kabara and placed under arrest, but she was able to bribe her guards to take her to Uganda, and thereby escaped. After hearing that she would be shot on sight if she returned to Zaire, she went to Rwanda, where she established a new research site that she named Karisoke, combining letters from nearby Mt. Karisimbi and Mt. Visoke (Fossey, 1983, p. 25).

Fossey's life in Karisoke was recounted by her in her book, *Gorillas in the Mist*, which was edited and rewritten with the goal of engendering popular support for the protection of the mountain gorillas (Mowat, 1987). Further material, such as a number of letters she wrote, and entries in her diary, was used to present the more personal aspects of Fossey's life in Mowat's book, *Woman in the Mists*. Other reports of her life and personality depict her as a bully (Fowler, 2019), or as a "a racist alcoholic who regarded 'her' gorillas as better than the African people who lived around them" (Varadarajan, 2002). Mowat's depiction of Fossey is more sympathetic, more balanced, and presents her as an imperfect human being, although he never actually met her in person. His book was derived from "thousands of her letters, her diaries and journals, and having listened to scores of people who knew her in life" (Mowat, 1987).

Worth considering, regarding Mowat's book, is a reader review and comment on Amazon.com that had this to say: "I enjoyed Mr. Mowatt's (sic) account of Dian's life and times with mountain gorillas. I was a research assistant there in 1974 and though he never personally interviewed me his account of my time there was very accurate" (Natoli-Rombach, 2012a), adding, "It gave an insight into Dian's thoughts and feelings that even those who worked with her could only guess at. We get to see the world of Dian Fossey as viewed through the mind and eyes of Dian herself," and "to say Dian was a troubled person at times is an understatement." The review continued with this sentiment:

> What I most appreciated is that even having personally known Dian I was able to get to know more of the whole Dian, the parts that came before and after I knew her. It gave me a better perspective on her life.

In another book about Fossey (Hayes, 1990), in which Natoli-Rombach is interviewed, Natoli-Rombach left the following comment:

> I truly liked Dian but hated her behavior at times that was often erratic and seemed to border on madness and even cruelty. One day she could be pleasant and the next day it seemed as if dark clouds descended on her head and everyone was her enemy. It was her behavior that probably posed the greatest risk to her research and reputation in her later years at Karisoke. In some ways her death helped prevent that and gave her a martyrdom status which history and the media capitalize on.
>
> *Natoli-Rombach, 2012b*

Cross-referencing with Natoli-Rombach's LinkedIn profile (Natoli-Rombach, n.d.) and comments on other web sites, which can be found using a Google search, I am reasonably confident that the person who posted the comments quoted above is indeed the same person mentioned in Mowat's book (Mowat, 1987, p. 377) and interviewed by Hayes (1990).

Cognitive Processes

Like Jane Goodall (Goodall, 2010), Fossey lacked any experience or formal training in field biology or as a researcher in any scientific endeavor. Aside from a brief visit at Gombe Stream Chimpanzee Reserve in Tanzania to observe Goodall's research process, Fossey had read, "virtually memorizing" (Fossey, 1984, p. 4) two books by George Schaller (1959; 1960) about his mountain gorilla research, as well as a book to teach herself Swahili. Upon her arrival at Kabala, she was able to employ a tracker who had worked for Schaller (Fossey, 1984, p. 3).

Otherwise, she had to learn everything about being a scientist on the job. Her own first solo attempt at tracking took her in the opposite direction the gorillas had gone. Perhaps she was able to succeed in becoming the world's foremost expert on mountain gorillas because, despite her lack of experience and training, she was passionately dedicated to the gorillas. This passion, as Polanyi (1957) has noted, had a "logical function" in terms of Fossey's research, and contributed "an indispensable element" to it (p. 114). For one, it drove Fossey's motivation, at least at the beginning of her research, as a passion to simply be in the presence of the gorillas, and for them to accept her. In this sense, her passion for her work as a researcher was inextricably intertwined with her passion for the gorillas themselves. Fossey derided others, especially graduate students, who came to Karisoke simply to get enough data for a dissertation thesis or academic publishing credentials. Only those who demonstrated a devotion to the gorillas that placed their well-being above all else did she seem to truly accept (Fossey, 1983; Mowat, 1987).

Fossey's research method was mostly observational, and her scientific contributions primarily descriptive. The data she gathered consisted of notes and drawings she made while observing gorilla behavior and interacting with gorillas, sound recordings of gorilla vocalizations and chest-beating, photographs and films of gorillas, samples of dung to be analyzed for parasites, and the bodies of deceased gorillas that were autopsied by a medical professional in an attempt to determine cause of death (Fossey, 1983; Mowat, 1987).

Fossey's cognitive style in research, her approach to understanding the gorillas, was primarily abductive. This was demonstrated by her putting to the test what she had read in Schaller's books about how to habituate gorillas, testing the hypothesis about the nature of gorillas presented by Schaller in the books. One might also describe her tracking of gorillas as based on abductive reasoning: A wide swath of crushed vegetation bent in the same direction, with "pools of diarrhetic dung" along the way, was a "flee trail" probably indicating that the gorillas had encountered either other gorillas or humans resulting in some sort of violent physical encounter (Fossey, 1983).

Unlike Polanyi's idea of connoisseurship (Polanyi, 1962, p. 54), which he described "as much an art of doing as it is an art of knowing" (p. 54), and "can be communicated only by example, not by precept" (p. 54), Fossey did not "go through a long course of experience under the guidance of a master" when it came to studying gorillas, at least not in person, and barely at all in any sense of the concept. Only when it came to tracking gorillas did she "rely on the transmission of skills and connoisseurship from master to apprentice" (p. 55). She was not brought into the field by an expert gorilla researcher who showed her the ropes, or perhaps the vines, so to speak. She had read a couple of books on the subject, spent a short time observing Jane Goodall's methods for studying chimpanzees, and was taught rudimentary tracking skills (although she usually relied on more seasoned trackers to lead her to gorillas). Aside from tracking, when it came to studying mountain gorillas, Fossey did not "rely on the transmission of skills and connoisseurship from master to apprentice" (p. 55). She developed these on her own, through trial and error.

Fossey's beginnings in the field research of gorillas got off to an ignominious start. She was accompanied to the research camp Kabara by a friend who could only stay for two days and who supervised the digging of a latrine and drainage ditches around Fossey's tent (Fossey, 1983, p. 6), and upon his leaving, Fossey was left alone (the only White person, but also the only woman) with two African staff who did not speak English. On the third day, she spotted a black, gorilla-sized creature apparently lying out in the sunshine, and watched it from an obscured position for over an hour, with binoculars, stopwatch, pen, and notebook at the ready. When she moved closer, the animal crept away into the bushes. Later on, she discovered it had been an old forest hog that had sought out a solitary place to lie down and die in (Fossey, 1983, p. 8). Eventually, she became more adept at finding gorilla groups to observe, especially with the help of an experienced tracker, and this seemed to be the main activity of her research: spending hours hiking through the brush to find gorillas, and then even more hours observing them, in every kind of weather (Fossey, 1983).

Once Fossey habituated gorillas to her presence, though, encounters with those gorillas became much more interactive. Habituation was necessary in order to be able to observe the gorillas up close; attempting to remain undetected by them necessarily kept her at a distance, and when they did detect her they would flee (Fossey, 1983). In order to habituate them, Fossey had to use deductive reasoning to determine how best to "put the gorillas at ease," which she did "by imitating regular activities like scratching and feeding, and copying their contentment vocalizations" (Dian Fossey Gorilla Fund, n.d.; Fossey, 1983), behaviors Fossey had observed from afar.

> She also came to depend on the gorillas' natural curiosity in the habituation process. While walking or standing upright increased their apprehension, she was able to get quite close when she "knuckle-walked." She would also chew on celery when she was near the groups, to draw them even closer to her.
>
> *Dian Fossey Gorilla Fund, n.d.*

However, at the beginning, she made a few mistakes, such as mimicking the gorillas' chest-beating by slapping her thighs with an open hand, which definitely got the attention of the gorillas, but sent the wrong message, as gorilla chest-beating is a "signal for excitement or alarm," (Fossey, 1983, p. 13). Thus, there was a lot of trial-and-error learning on Fossey's part.

Polanyi's idea of a "mutual correlation between the personal and the universal within the commitment situation" (1962, p. 302) applies to Fossey's approach to her research (most importantly with regard to habituation of the mountain gorillas). According to Polanyi, "commitment is a personal choice, seeking, and eventually accepting, something believed (both by the person incurring the commitment and the writer describing it) to be impersonally given" (p. 302). Polanyi elaborates that one "can speak of facts, knowledge, reality, proof, etc." as "proper designations for commitment targets" but only if one is committed to them as such for universal validity (p. 302).

Fossey's approach to habituation had an "impersonal status" (p. 302) as having been established, if not by science then by at least a scientist, in this case Schaller. Her adoption of his approach was an intensely personal commitment: being wrong could mean suffering a vicious attack by a 400-lb silverback gorilla. Furthermore, Polanyi's idea also explains, at least in some sense, Fossey's dedication to antipoaching activities.

As Polanyi writes (1962), so called "'actual facts' are accredited facts, as seen within the commitment situation, while subjective beliefs are convictions accrediting these facts as seen noncommittally by someone not sharing them" (p. 304). Fossey's apparent obsession with antipoaching activities grew from her commitment situation. Yet, despite her interactions with gorillas, she did not believe that her presence overtly affected gorilla behavior to the point where they behaved unnaturally. Humans were already part of their environment in the form of poachers, and local peoples who farmed and ranched in the foothills of the mountains where the gorillas lived. Also, as mentioned before, habituation allowed her to observe gorillas up close without putting them ill-at-ease, allowing them to behave as if she were merely a normal part of the environment. There was a precedence of sorts, but also a difference, in the research of Jane Goodall, in which chimpanzees were lured to her camp by the presence of bananas (Goodall, 2010).[1]

One might say that poaching activities, as well as interactions with African farmers, ranchers, hunters, were regarded by Fossey as intrusive, an aberration in the environment the gorillas inhabited. This may be interpreted as essentially racist, but it also may be interpreted as because she viewed these interactions as universally harmful to the gorillas, destructive of their habitat, and disruptive of their social cohesion. Undermining the racism interpretation, however, it must be noted that Fossey held the same view toward White tourists and zoo officials whom she felt had the same generally negative impact on the gorillas. She also found herself, on several occasions, put into the situation of caring for infant gorillas that had been taken by others from the wild, either to sell to tourists or to European zoos, and had usually suffered from the trauma surrounding their being taken, as well as malnutrition and the ongoing stress of captivity (Fossey, 1983; Mowat, 1987). She had a low opinion of anyone who would do anything to harm the gorillas.

Fossey used inductive reasoning, although not to the extent of developing formal theories, as such. She developed an understanding of what would happen when a zoo requested that wild young gorillas be obtained for display because she had observed gorillas defending their group against incursions by rival males (Fossey, 1983). The only way to successfully capture a baby gorilla was to kill the defending adults, which often included the silverback. This would likely result in the dissolution of the group if there wasn't an older, related male to take over, forcing females to join other groups. Any unweaned young would be killed by silverbacks of the new groups to force their mothers into estrous. Thus, in order to acquire two young gorillas for display in a zoo, at least two adult gorillas would likely be killed (the mother and at least one defending male), and possibly more would die soon after through infanticide (Fossey, 1983; Mowat, 1987). Fossey generalized this behavior to all mountain gorillas in the context of Darwinian evolution, and this knowledge contributed to her passion for antipoaching activities.

The concept of scientific detachment cannot be applied to Fossey and her research process. Before her, Goodall gave her chimpanzee research subjects names, which went against the then-standard practice of assigning numbers to the animals being researched in order to maintain that detachment, although this was not as radical a break, as others had done so before her (Benson, 2016). Goodall's method had the added benefit of making it easier for her to differentiate between individual chimpanzees (Goodall, 2010, p. 32). Fossey did the same. Far from being detached, Fossey regarded some of the gorillas she studied as dear friends, and delighted in being accepted by them as nonthreatening. In fact, much of her book focuses on her relationships with different gorillas, especially Digit (Fossey, 1983). Digit would often come to her during her long hours of observing, investigate her gear, flop over onto his back, and "invited play," Dian, wrote, stating, "(a)t such times, I fear, my scientific detachment dissolved" (p. 182).

Social Processes and Fossey's Research

The larger cultural and political climate of the area and time in which Fossey did her research exacted a huge influence on her work, but this could be said of any researcher in any place at any time. It is the specifics that matter: The forces at play included recently postcolonial independent nations dealing with multiple insurrections, rebellions, and attempted coups, all amid the even larger setting of the Cold War in which both sides struggled for influence in Africa, supporting either governments or rebels as proxies (Atomic Heritage Foundation, 2018). Furthermore, in the realm of science, women were still an anomaly, hence the great deal of publicity accorded to Leakey and his "Trimates," Goodall, Fossey, and Galdikas (Mowat, 1987). Had these three been men, one wonders if there would have been nearly the same level of interest in, and critical scrutiny of, their work and, especially, their personal lives.

The local social processes of Fossey's research involved both human and non-human, primarily but not exclusively gorilla, social elements. Louis Leakey was her conduit to the opportunity to research the mountain gorillas, and it was he who arranged her funding, and purchased a Land Rover for her (Fossey, 1983). Eventually, National Geographic funded her, and after Leakey died, Fossey had to solicit funding herself. Much of the funding success seemed to owe much to Jane Goodall's success with chimpanzees (Mowat, 1987). Fossey's role with regard to funding sources obligated her to write articles and participate in the creation of documentaries, to give talks at conferences and other events, and concede to interviews with the media (Fossey, 1983; Mowat, 1987).

Aside from funding sources, Fossey's most important social connections were the people who lived in Africa, both of European decent and local peoples.

She depended on connections with Whites for her most intimate interpersonal relationships; they tended to be wealthy, or at least wealthy compared to the Africans she associated with (outside of government officials). When it came to local peoples, her relationships were a mixed bag. On the one hand, she was the sworn enemy of poachers and the persistent thorn in the side of farmers, whose cattle she would kill or maim or hold hostage if they ranged illegally into areas reserved for wildlife. On the other hand, she greatly respected her top trackers, and the park guards who shared her devotion to preserving the mountain gorillas and their habitat earned her admiration. Even though she understood that for local peoples the designation of a boundary demarcating where it was okay to hunt or graze cattle was essentially meaningless to them because they'd used the land for their own purposes for generations, she prioritized the gorillas because she knew them to be on the brink of extinction, and she cared about individual gorillas (Fossey, 1983; Mowat, 1987). And, as mentioned before, she cared about people who cared about gorillas. She had little use for those who did not, regarded as adversaries those who undermined her control of Karisoke, and regarded as enemies those who endangered the lives of gorillas.

According to some, Fossey was a racist, imperialist, privileged White woman, who referred to Africans as "wogs" and tortured poachers (Mowat, 1987; Varadarajan, 2002; Rodrigues, 2019). While not necessarily completely inaccurate, this is a one-sided description of her which ignores evidence that she also greatly respected and worked well with many Africans. For example, one of the Rwandan trackers who worked with her was unusually tall, with unusually big feet that no mass-produced boots could fit. Fossey made boots for him herself, using tire rubber, which were a poor substitute but served their purpose until someone in the United States had boots custom-made for him after reading about the situation. The point here is that, this man with the oversized feet was dedicated to active conservation and believed that the mountain gorillas were a national treasure worth protecting, going out to destroy snares and run off poachers even without proper footwear, and therefore earned Fossey's respect and admiration (Mowat, 1987). He cared about the gorillas, and therefore Fossey cared about him, as she did other locals who demonstrated a similar attitude toward the gorillas (Fossey, 1983; Mowat, 1987).

Fossey's social environment also included nonlocal visitors to the research facility. These were mainly graduate students, who could stay for days or years (Fossey, 1983; Mowat, 1987), but there were also visiting photographers and films crew, as well as (mostly White) tourists. She had zero respect for tourists, whose presence disrupted not only her life and research, but the lives of the gorillas (Fossey, 1983; Mowat, 1987). Tourists could be demanding and insensitive to the needs of the gorillas. Once, Fossey fired a gun over the heads of a group of tourists, who then attempted to have her charged by the Rwandan authorities with attempted murder. However, the president of Rwanda supported her research and had the charge dismissed (Mowat, 1987). She tolerated visiting photographers and film crews, if they behaved properly, and even had an affair with a photographer (Fossey, 1983; Mowat, 1987).

Some of her most rewarding relationships, and some of her most contentious, were with graduate students who came to do their own research. There was underlying tension with those whom Fossey deemed were there simply to advance their own careers as opposed to being dedicated to the protection of the gorillas (Fossey, 1983; Mowat, 1987). Some of the most damning "evidence" of Fossey's bad behavior came about due to reports from graduate students who worked with her, portraying her as an alcoholic, gun-wielding, hysterical racist obsessed with destroying poacher traps while impeding the progress of legitimate research (Fossey, 1983; Mowat, 1987). She was often lonely, developed romantic attachments, sometimes ones-sided, with males who visited Karisoke, as well as with a married local doctor, and when relationships ended, she did not take it well, resorting to drinking alcohol as a way to cope (Fossey, 1983; Mowat, 1987). Female students could be perceived as rivals for the attentions of male students, and therefore ill-treated (Fossey, 1983; Mowat, 1987).

Some of her graduate students attempted to have her removed as director of the Karisoke research center, waging a war of letters written to National Geographic and others. Fundamental ideological differences with regard to how best to fund and conduct scientific research were at play (Fossey, 1983; Mowat, 1987). These involved two major issues: whether or not to promote tourism in the area as a way to generate funds to pay for research, and whether or not researchers should be involved at all in antipoaching activities. In the epilogue to her book, however, Fossey (1983) did seem to relax her stance regarding tourism, writing that "tourism, if properly directed, might well prove profitable on a nationwide basis and thus compel the one-to-one reapers of wildlife proceeds to give way to the rule of the majority" (p. 241).

Conclusion

Fossey cannot be regarded separately from her times, and her research cannot be regarded separately from her personality, values, and passions. In some ways, she was ahead of her time as a researcher. She anticipated the need for active, hands-on conservation, and understood that mountain gorillas would not survive without the involvement of humans. However, also as a product of her times, inserted into a barely postcolonial central Africa where Whites were still transitioning from the status of those in power, she was permitted to behave in abhorrent fashion toward African poachers and farmers.

Fossey understood that she had two jobs: research and conservation, and neither could succeed without the other. If the gorillas were not protected, there would nothing to conduct research on, and research was necessary to understand the needs of gorillas and develop better means of protecting them. While the research, especially being in the field with the mountain gorillas, was her passion, it was conservation, especially active conservation represented by antipoaching activities, that became her obsession. As Schaller stated in an interview (2006):

> Conservation has nothing to do with animals, it has to do with economics, with society, with politics. Suddenly, you find most of your time is spent on things you don't really want to do because that's not how you started out. But in the end, the conservation is more important.

Without the research, conservation is based on guesswork about what is best for the species being protected. Without conservation, there may nothing to study, at least nothing alive in the wild.

QUESTIONS FOR DISCUSSION

1. Is formal training and education necessary for one to become an effective scientist?
2. In the case of field biology, especially, what kind of relationship should the scientist have with her subject(s) of study?
3. For a scientist studying endangered species, can, or should conservation activism and research be unified activities? Or should they remain completely separate?
4. Should a field biologist be concerned with the well-being of specific animals among the population they are studying to the point of intervening to benefit an animal? Are there instances where this should, or should not, happen?
5. Should a field biologist ever conduct an activity that may result in harm to an animal being studied? Under what circumstances might this be allowable?

Note

1 "Not until Hugo and I had actually left the Gombe Stream did we realize that during the year we had made one grave mistake." They had encouraged touching (grooming and tickling behaviors) between chimpanzees and themselves. They realized that a young male chimpanzee was much stronger than a human, and they "realized the foolishness" of this behavior. A chimpanzee who realized he was much stronger than a human researcher, and had no fear of that researcher, "would become dangerous." Furthermore, "repeated contact with a wild animal is bound to affect its behavior" (Goodall, 2010, pp. 136–137). The last quote can be applied to Fossey's habituation of mountain gorillas, although perhaps due to the nature of mountain gorillas, or their social structure, or the fact that Fossey did not lure them to her camp with irresistible treats like bananas, or her ability to put them at ease, or just good luck, (or some combination of these), the affect

it seemed to have on the mountain gorillas is what made it easier for her to study them in close physical proximity with them.

References

Atomic Heritage Foundation. (2018, August 24). Proxy wars during the Cold War: Africa. *Cold War History.* www.atomicheritage.org/history/proxy-wars-during-cold-war-africa

Benson, E.S. (2016,) Naming the ethological subject. *Science in Context.* *29*(1), 107–128. doi: 10.1017/S026988971500040X. PMID: 26903374

Dian Fossey Gorilla Fund International. (n.d.). Dian Fossey bio. https://gorillafund.org/who-we-are/dian-fossey/dian-fossey-bio/

Fossey, D. (1972). Vocalizations of the mountain gorilla. *Animal Behavior, 20,* 36–53.

Fossey, D. (1974). Observations on the home range of one group of mountain gorilla. *Animal Behavior, 22,* 568–581.

Fossey, D. (1977). Feeding ecology of free-ranging mountain gorillas. In T. H. Clutton-Brock (Ed.), *Primate ecology: Studies of feeding and ranging behaviour in lemurs, monkeys, and apes* (pp. 415–447). Menlo Park, CA: Academic Press.

Fossey, D. (1979). Development of the mountain gorilla (*Gorilla gorilla beringei*): The first thirty-six months. In D. A. Hamburg & E. R. McCown (Eds.), *The great apes* (pp. 139–184). Menlo Park, CA: Benjamin-Cummings.

Fossey, D. (1982). Reproduction among free-living gorillas. *The American Journal of Primatology, 3*(21) (supplement 1), 97–104.

Fossey, D. (1983). *Gorillas in the mist.* New York: Mariner Books.

Fossey, D. (1984). Infanticide in mountain gorillas (*Gorilla gorilla beringei*). In G. Hausfater & S. B. Hrdy (Eds.), *Infanticide: Comparative and evolutionary perspectives* (pp. 217–235). New York: Aldine.

Giere, R. N. (1979). *Understanding scientific reasoning.* New York: Holt McDougal.

Goodall, J. (2010). *In the shadow of man.* New York: Mariner.

Hayes, H. T. P. (1990). *The dark romance of Dian Fossey.* New York: Simon & Schuster.

Jane Goodall Institute. (n.d.). Our story: Timeline. www.janegoodall.org/our-story/timeline/

Leakey Foundation. (n.d.). The Leakey family: Louis Seymour Bazett Leakey. https://leakeyfoundation.org/about/the-leakey-family/

Mowat, F. (1987). *Woman in the mists: The story of Dian Fossey and the mountain gorillas of Africa.* New York: Warner Books.

Natoli-Rombach, R. (2012a, August 3). Brought back lots of memories…. [Comment on the book "Woman in the Mists"]. *Amazon.com.* www.amazon.com/gp/customer-reviews/R6ORHOILV0VLS/ref=cm_cr_getr_d_r vw_ttl?ie=UTF8&ASIN=0446387207

Natoli-Rombach, R. (2012b, August 3). Mr. Hayes interview me for this book in 1987… [Comment on the book "The Dark Romance of Dian Fossey"]. *Amazon.com.* www.amazon.com/gp/customer-reviews/RP2PXC4HBSDI/ref=cm_cr_dp_d_rvw   _ttl?ie=UTF8&ASIN=0671633392

Natoli-Romach, R. (n.d.). LinkedIn profile. *LinkedIn.com.* www.linkedin.com/in/richard-natoli-rombach-2403b758/

O'Doherty, K. C., Osbeck, L. M., Schraube, E., & Yen, J. (2019). *Psychological studies of science and technology.* Cham: Palgrave.

Orangutan Foundation International (n.d.). Dr. Biruté Mary Galdikas: Dr. Galdikas biography. https://orangutan.org/about/dr-birute-mary-galdikas/

Paulson, S. (2017, December 2). The women who revolutionized primatology. *To the best of our knowledge*. www.ttbook.org/interview/women-who-revolutionized-primatology

Polanyi, M. (1962). *Personal knowledge: Towards a post-critical philosophy*. Chicago, IL: University of Chicago Press.

Rodrigues, M. A. (2019, September 20). It's time to stop lionizing Dian Fossey as a conservation hero. *Lady Science*. www.ladyscience.com/ideas/time-to-stop-lionizing-dian-fossey-conservation

Schaller, G. B. (1963). *Mountain gorilla: Ecology and behaviour*. Chicago, IL: University of Chicago Press.

Schaller, G. B. (1964). *The year of the gorilla*. Chicago, IL: University of Chicago Press.

Schaller, G. B. (2006, August). Interviews: Dr. George B. Schaller. *Sanctuary Asia*. www.sanctuaryasia.com/interviews/gbschaller.php

Varadarajan, T. (2002, May 4). Giants of the jungle. *Wall Street Journal*. www.wsj.com/articles/SB1015200880517583680

A COMMENTARY ON THE TEACHING CASE STUDIES IN *PERSON-CENTERED STUDIES IN PSYCHOLOGY OF SCIENCE*

Examining the Acting Person

Ronald B. Miller

As one who has spent most of his academic career in psychology questioning whether the field's mainstream experimental research methodology was adequate for the study of complex human emotions, motivations, and real-world interpersonal relationships, I am heartened to discover the progress that Dr. Lisa Osbeck and colleagues have made in articulating both the theoretical framework and the research methodology for a *person-centered psychology of science*. As Dr. Osbeck has indicated in the introductory chapter to this volume, a concerted effort is underway to reclaim the perspective of the humanistic, phenomenological, and narrative perspectives of mid-20th century psychologists Abraham Maslow, Carl Rogers, physical scientist Michael Polanyi, when describing and accounting for the actual work of renowned scientists.

The central question is, of course, how to actually carry out research studies that accurately portray the human beliefs and actions that are central to the pursuit of successful scientific research. The central answer to this question is to conduct multilayered narrative interviews with various members of a given research team, observe the research team meetings, and record group discussions for later qualitative analysis (see Osbeck and Nersessian, 2012). One can imagine that many research labs would be wary of opening their inner workings to person-centered psychology of science researchers (if only for fear their proprietary research methods or findings will be revealed to competitor laboratories).

In a typical graduate course on the psychology of science participation in such a person-centered psychology of science study by a group of students would typically be beyond the realm of possibility. Yet, it is very important to have an experiential component that requires the graduate student the opportunity to begin to tackle the myriad psychosocial factors that might influence or alter the scientist's

DOI: 10.4324/9781003276692-11

perception of the scientific question that is being investigated. Professor Osbeck's pedagogical solution to this dilemma is as innovative as her conceptual analysis of the person-centered psychology of science. She assigned her graduate students to choose a scientist of interest and then identify through well-respected biographies and/or autobiographies of the scientist in question, how this scientist's life story and their personal actions and beliefs, contributed to their scientific advance. These case studies are categorized as "teaching case studies" in that the graduate student can benefit greatly by searching in the biographies for personal information about the scientist that may provide insights into the cognitive, social, emotional, familial, or personality traits that contribute to the scientific discoveries for which they are credited.

However, since the accuracy or validity of this biographical information is assumed, but not directly verifiable by the author of the teaching case study, the conclusions drawn in the teaching case study can only be conditional on the care taken by the original biographer to present reliable and valid information. This limitation of the individual teaching case studies does not, however, restrict the reader from considering the commonalities across the various "teaching case studies," especially when these case studies span the lives of scientists who worked in quite varied fields, across different centuries, continents, and disciplines of inquiry. Such parallels and commonalities or sharp differences across different cases suggest avenues to pursue in more formal case studies in the psychology of science. It is exactly such common themes across the disparate "teaching case studies" that I found most compelling and convincing as to the strength of Osbeck's account of a person-centered psychology of science.

Themes across Case Studies

The Search for Meaning in Scientific Discovery

From M.A. Khalid's, Johannes Kepler, born in the 16th century and building the foundation of modern astronomy, to S.L. Antczak's Dian Fossey conducting groundbreaking 20th century studies of great apes in Africa, we see eight scientists in this book whose biographies document their search for meaning and purpose in life through their contributions to science. These scientists did this by writing about a reconciliation between reason and Christian faith (Kepler and Y. Raza's Poincaré), communing with nature as they studied it (M.V. Steder's Goethe and P. Schillemat's Carson), contributing to the development of their science as well the greater good of their communities (A. Asad's Mendeleev, G.F. Crowe's Boas, and R. Hopkins' Nash) or promoting humanity's self-understanding (Nash's game theory, and Boas' creation of anthropology as an antidote to racial and religious prejudices). One might say that each of them struggled with reconciling their own brilliant insights into the workings of our world with the received wisdom of their time, and how that received wisdom had impacted their own life.

The Role of Intuitive Unconscious Creative Processes in the Development of Scientific Knowledge

When students begin their study of the sciences, an attempt is made to offer research methods paradigms that generate empirical and quantitative data that, when sufficiently collected, will lead to inductive generalizations about the laws of nature. These laws are seen as open to revision or the discovery of additional laws of nature as more research data are carefully collected. The role of creative problem-solving or inspired insights in directing research programs is viewed generally as peripheral to the scientific method. Yet, in every case study presented in this volume we see researchers drawn to areas of study by a compelling attraction to the subject matter, and then when frustrated by the lack of progress in understanding the phenomena in question, a solution comes to mind seemingly out of the blue or in a dream. Nash even claimed that solutions to mathematical equations appeared to him as visual images that were as vivid as his psychotic hallucinations. Poincaré could only solve difficult mathematical problems when he ceased from his mathematical labors and walked around thinking of nothing in particular. Mendeleev was unable to resolve a rational table of chemical elements until he was reminded of his favorite card game "Patience" and the way cards had to be arranged in a specific pattern of rows and columns.

Antczak (this volume, pp.17) states that Fossey habituated the male gorilla "Digit" so well to her presence that during her long hours of observing him *in his natural habitat* "he would investigate her gear, flop over onto his back and invited play." While this differs from the above unexpected sudden intuitive solutions to otherwise unsolvable problems by research scientists, it strikes this reader that here, too, Fossey discovered something about "disarming" the much feared African male gorillas that others had never reported.

The Moral Dimension of the Psychology of Science

By identifying the pursuit of science by renowned scientists with the search for meaning in life, we find ourselves recognizing that even the pure sciences are pursued by human beings who want answers to life's mysteries and miseries. When we as a society prioritize the funding of this or that research proposal, we are expressing what problems are to be addressed and whose problems are to hopefully solved. When these decisions are made by public funding sources, private businesses, or private foundations, the decisions reflect the priorities and values of the community or those private entities. When scientists decide to put their intellect and resources to work on a given project, they are implicitly or explicitly endorsing the importance of the project and its value to some segment of the society.

To the extent that one portion of the population benefits more than others in such "scientific" decisions, the decision is a moral one for the individuals who participate, and perhaps a social or political one for the culture as a whole. Of course, there are many scientific or technical questions as well about how to do the scientific research in question, but whether to do the scientific project is a fundamentally moral/ethical decision. As a society, the United States has typically worked very hard to present research spending priorities to the public as matters to be left to the technically competent experts, ignoring the moral implications of doing so.

The psychology of science has the opportunity to reintroduce into the discussion of the validity of scientific research designs and findings the moral values implicit in the scientific question being studied, the manner in which the study was conducted, its findings, and how those findings are reported and interpreted by the researchers. Only then will we be able to determine the moral, social, and political value to our society of the research findings. In the absence of such attention to the moral dimension of the psychology of science, we will contribute to the further corruption of what passes for valid scientific progress in the world scientific community.

This critical issue was addressed forcefully by Ad Lagendijk (2005) in an essay *Nature*, "Pushing for Power." At the time he was the group leader at the FOM-Institute for Atomic and Molecular Physics, Amsterdam and university professor at the University of Amsterdam, Netherlands. Here are a few of the key passages:

> Tales of brilliant scientists and their heroic discoveries can overshadow the true nature of scientific communities, which are often dominated by battles for power and success….The primitive value system—tallies of publications, citations, and patents—now used in science is the cause of this obsession with power rather than with curiosity and scientific progress. But does this system give us, in the long run, the best value for our money. I doubt it.
>
> *p. 429*

One might think that such competition would sharpen the thinking and increase the value of new scientific discoveries, but that is not the case. Lagendijk notes that at scientific conferences in his area of expertise, atomic and molecular physics, he sees male physicists (there are very few female ones) fighting to defend their scientific claims that are essentially *unimportant findings*. Clearly, the sciences are in trouble and they need a new grounding. A person-centered psychology of science offers a critical vantage point from which to examine the decaying physicalist paradigm Lagendijk has so forcefully and clearly identified.

References

Lagendijk, A. (2005). Pushing for power. *Nature, 438*(7067), 429–429.

Osbeck, L. M., & Nersessian, N. J. (2012). The acting person in science practice. In R. W. Proctor & E. J. Capaldi (Eds.). *Psychology of science: Implicit and explicit processes* (89–111). New York: Oxford University Press.

INDEX